Goat Guru: Navigating the Basics of Raising Happy Hooves

A Comprehensive Manual for Novice Herdsmen

Alex Turner

Table of Contents

INTRODUCTION

Welcome to "Goat Guru: Navigating the Basics of Raising Happy Hooves - A Comprehensive Manual for Novice Herdsmen." Suppose you've ever dreamed of embarking on the rewarding journey of raising goats or find yourself at the threshold of becoming a novice herder. In that case, this comprehensive guide is your key to unlocking the secrets of successful goat farming.

In the heart of this e-book lies a wealth of knowledge designed to empower you with the skills and understanding needed to raise a content and a thriving herd of goats. Whether you're interested in dairy, meat, fiber, or simply seeking the joy of companionship these charming creatures bring, "Goat Guru" is your go-to resource.

Our journey begins with an exploration of the fascinating world of goats. Understanding the diverse breeds and their characteristics and deciphering the intricacies of goat behavior lays the foundation for a strong connection with your herd. With this knowledge, we delve into the practical aspects, guiding you through the essential steps of setting up your goat farm - from planning and building shelters to creating an environment that caters to their unique needs.

As you progress through the chapters, you'll learn how to select the suitable breeds for your goals, create a balanced nutrition plan, and navigate the intricacies of goat health and veterinary care. Breeding, kid care, and training techniques are demystified, ensuring you can confidently handle every aspect of goat farming.

"Goat Guru" is not just a manual; it's a companion on your journey to becoming a skilled herdsman. We cover the basics, explore profitable ventures and marketing strategies, and troubleshoot common challenges. As you immerse yourself in goat farming, this e-book becomes a trusted guide, offering insights, tips, and practical advice that will elevate your goat-raising experience. So, let's embark on this adventure together, nurturing happy hooves and cultivating a thriving goat community.

CHAPTER I

Understanding Goats

The Fascinating World of Goats

The world of goats is a captivating and diverse realm. These remarkable creatures have carved out a niche as companions, milk, meat, and fiber providers, and even as essential players in sustainable agriculture. Understanding the fascinating facets of the goat kingdom involves unraveling their rich history, exploring the vast array of breeds, and delving into the intricacies of their behavior.

Goats, often considered one of the oldest domesticated animals, have been intertwined with human history for thousands of years. From the wild bezoar ibexes of Asia and Europe, goats accompanied humanity from nomadic lifestyles to settled agriculture. Ancient civilizations valued goats for their meat and milk and their practical roles in clearing land, providing leather, and offering companionship. The ubiquity of goats in various cultures is reflected in mythology and religious symbolism, where they often represent symbols of fertility, sacrifice, and prosperity.

As we navigate the expansive world of goats, understanding the diverse breeds that inhabit it becomes paramount. Goats come in various shapes, sizes, and colors, each adapted to specific environments and purposes. From the small and hardy Nigerian Dwarf, prized for its milk production, to the large and imposing Boer, bred for meat, and the luxurious Angora, renowned for its silky fiber, the spectrum of goat breeds is as vast as it is intriguing. The distinct characteristics of each breed extend beyond mere physical appearance; they encompass temperamental traits, dietary preferences, and adaptability to different climates.

Behavioral insights into the world of goats add another layer of fascination. Goats, often associated with mischief and stubbornness, possess an intelligence that sets them apart. Their social structure, communication methods, and problem-solving abilities contribute to their reputation as engaging and charismatic animals. Understanding the nuances of goat language, from the various bleats and calls to body language, is critical to establishing a harmonious relationship with these creatures. Goats thrive on companionship, and their intricate social hierarchies and bonds with humans make them not just livestock but valued members of a herd.

Beyond their role as companions and sentient beings, goats play a crucial part in sustainable agriculture. Their voracious appetites and ability to thrive in diverse landscapes position them as efficient weed controllers and land clearers. The practice of goat grazing for land management has gained popularity in addressing issues such as invasive plant species and fire prevention. This symbiotic relationship between goats and the environment underscores their significance beyond traditional farming.

The nutritional aspects of goat husbandry also contribute to the allure of these creatures. Goats are renowned for their versatility in adapting to various climates and dietary conditions. Whether foraging on sparse vegetation in arid regions or thriving on lush pastures, goats exhibit a remarkable ability to extract nutrients from diverse sources. Crafting a balanced diet that caters to their specific needs is essential for their health and productivity, emphasizing the necessity for herders to comprehend the nutritional requirements of their chosen breed.

As we explore the fascinating world of goats, it becomes evident that their contribution to human welfare extends beyond the immediate provision of milk, meat, and fiber. Goats have become symbols of sustainable agriculture, embodying resilience, adaptability, and resourcefulness. In an era where sustainability and environmental consciousness are paramount, the goat's role as a custodian of the land underscores its relevance in shaping the future of farming practices.

In conclusion, the captivating world of goats encompasses not only the practical aspects of farming but also delves into the historical, cultural, and ecological dimensions. From their ancient domestication to their diverse breeds, intelligent behavior, and vital roles in sustainable agriculture, goats embody a complex tapestry that continues to unfold. To fully appreciate and navigate this intricate world, aspiring herders and enthusiasts alike must embrace the multifaceted nature of these remarkable creatures, fostering a deep respect for their historical significance and their potential to contribute to a sustainable and harmonious coexistence with humans.

Different Breeds and Their Characteristics

The world of goats is a tapestry woven with diverse breeds, each contributing a unique set of characteristics shaped by centuries of selective breeding and adaptation to specific environments. Understanding the intricacies of different goat breeds is not merely an exercise in classification; it is a journey into the nuanced traits that define these animals, from their physical attributes to their temperamental dispositions. As we explore the vast array of goat breeds, we embark on a captivating odyssey that unveils the multifaceted nature of these creatures.

One of the most iconic goat breeds is the Nubian, known for its distinctive appearance with long, pendulous ears and a Roman nose. Originating in Africa, Nubians have adapted well to various climates and are prized for their high milk production, making them a popular choice among dairy enthusiasts. Beyond their utility, Nubians are characterized by a friendly and outgoing temperament, making them ideal companions for those seeking not just productivity but also a more personal connection with their herd.

In contrast, the Alpine goat, hailing from the Swiss Alps, is renowned for its exceptional milk production and adaptability to mountainous terrains. With various coat colors and patterns, Alpines exhibit a robust and hardy constitution, making them well-suited for a range of climates. The Alpine's versatile nature extends to its role in dairy and meat production, making it a favored choice for those seeking a dual-purpose breed that excels in multiple capacities.

The Boer goat, originally from South Africa, has gained global recognition for its impressive meat production. Boers are characterized by a distinctive white body and a red-brown head, and they possess a fast growth rate, making them efficient meat producers. Their hardiness and adaptability have led to their widespread adoption in various climates, cementing their status as a leading choice for meat-focused goat farming.

The Nigerian Dwarf, a diminutive breed with roots in West Africa, has captured the hearts of many with its small stature and endearing personality. Despite their size, Nigerian Dwarfs are prolific milk producers, making them an excellent choice for those with limited space. Their friendly and sociable nature further enhances their appeal, especially for small-scale farmers or those seeking goats as pets.

For enthusiasts interested in fiber production, the Angora goat emerges as a prime candidate. Originating in Turkey, Angoras are recognized for their luxurious and lustrous mohair, a prized fiber with exceptional sheen and warmth. These goats require meticulous grooming and care to maintain their fiber quality, attracting those who appreciate the artistry of fiber arts and the unique challenges of raising these elegant animals.

The Saanen, a Swiss dairy breed, stands out with its pure white coat and high milk production. Renowned as one of the best dairy breeds, Saanens are often favored by commercial dairy farmers for their consistent milk yield and docile temperament. Their adaptability to different management systems and climates further contributes to their popularity within the dairy industry.

Another dairy powerhouse is the LaMancha, recognized for its distinct ear characteristics—either very small "gopher" ears or virtually nonexistent "elf" ears. Originating in the United States, LaManchas have proven their worth in milk production and exhibit a calm and gentle demeanor. Their adaptability to various climates makes them suitable for a wide range of farming environments.

The Pygmy goat, a breed originating from West Africa, has become a beloved choice for those seeking compact and playful companions. Known for their small stature and playful antics, Pygmies are often kept as pets and are particularly popular in urban and suburban settings. While they may not be known for prolific milk or meat production, Pygmies contribute to the goat world with their charm and suitability for small-scale farming.

The Cashmere goat, originating from the Himalayan region, is valued for its fine undercoat, which is harvested to produce luxurious cashmere fiber. Characterized by a soft and insulating coat, Cashmeres thrive in colder climates and have found a niche market among those passionate about high-quality fiber production. Their unique contribution to the textile industry underscores the diverse roles that goats play in meeting human needs beyond traditional agriculture.

The Boer, Kiko, and Spanish goats stand out as robust meat producers, each with its own set of characteristics suited to specific environments. Boers, originating in South Africa, have gained global popularity for their fast growth rates and high-quality meat. Kikos, developed in New Zealand, exhibit exceptional foraging abilities and adaptability, making them well-suited for meat production in diverse landscapes. Spanish goats, descended from the goats brought by Spanish explorers to North America, are known for their hardiness and ability to thrive in challenging conditions, making them valuable contributors to sustainable meat production.

As we traverse the spectrum of goat breeds, it becomes evident that each has been meticulously shaped by the demands of its environment and the preferences of those who raised them. Beyond physical attributes, the temperament and behavior of each breed contribute to its suitability for specific roles within the agricultural landscape. From the hardworking and adaptable Boer to the elegant and fiber-producing Angora, each breed offers a unique story and set of characteristics that cater to the diverse needs and preferences of herders and farmers.

In conclusion, the fascinating world of goat breeds is a testament to the enduring partnership between humans and these remarkable animals. From the arid landscapes of Africa to the lush pastures of Switzerland, goats have adapted and evolved, becoming invaluable companions in our quest for sustenance, companionship, and sustainable agriculture. The characteristics that define each breed are not just the result of selective breeding but also a testament to the intricate dance between nature and nurture. As we continue to explore and appreciate the diversity of goat breeds, we gain not only a deeper understanding of these animals but also a profound appreciation for the role they play in shaping our agricultural practices and enriching our lives.

Behavioral Insights: Decoding Goat Language

Delving into the behavioral insights of goats unveils a complex and fascinating language that, once understood, forges a deeper connection between herder and herd. Goats, often portrayed as mischievous and headstrong, possess a rich tapestry of communication methods, ranging from vocalizations to intricate body language. Decoding this language becomes imperative for any herder seeking to nurture a harmonious relationship with their goats.

Central to goat communication is their vocal repertoire, a symphony of bleats, calls, and grunts that convey a myriad of emotions and needs. The distinct bleat of a mother goat searching for her kid, the excited calls during feeding time, or the plaintive cry of a goat in distress are all part of this auditory language. Understanding the nuances of these vocalizations allows herders to respond appropriately to the needs of their goats, fostering a sense of trust and security within the herd.

Equally significant is the body language exhibited by goats, a silent but eloquent form of communication. A raised tail, for instance, signals excitement or anticipation, while a lowered head may indicate submission or contentment. The complex hierarchy within a goat herd is often expressed through subtle cues such as ear position, head butting, and the stamping of hooves. Observing and interpreting these non-verbal signals enables herders to navigate the social dynamics of the herd and intervene when necessary to maintain a peaceful equilibrium.

The concept of dominance and submission is intrinsic to goat behavior, establishing a hierarchical order crucial for maintaining order within the herd. Head butting and pushing are standard displays of dominance, as goats assert their positions within the social structure. Understanding these interactions is paramount for herders to intervene when necessary, ensuring the well-being of each goat and preventing the emergence of disruptive behaviors that may compromise the overall harmony of the herd.

Goats are social animals, and their need for companionship and interaction is integral to their well-being. Separation anxiety and stress can manifest when goats are isolated from their herd, emphasizing the importance of considering the social dynamics when managing and housing these animals. Whether establishing a pecking order, engaging in play, or seeking comfort in each other's company, goats constantly communicate through their actions, reinforcing the interconnectedness of the herd.

In the realm of reproduction, goat behavior takes on a distinctive character. During the breeding season, bucks engage in elaborate displays of courtship, including vocalizations, headbutting, and even marking their territory with scent glands. Recognizing these behaviors is crucial for successful breeding management, ensuring that does are bred at the optimal time for successful pregnancies. Additionally, understanding the signs of heat and estrus in does allows herders to plan and manage breeding seasons effectively.

Goat behavior also extends to their response to environmental stimuli and perceived threats. Goats are known for their keen awareness of their surroundings, and their flight-or-fight instincts are finely tuned. A sudden noise, the presence of a predator, or an unfamiliar object can trigger a rapid response, with goats either fleeing or standing their ground in defense. Herders must be attuned to these reactions to create a secure environment that minimizes stress and anxiety within the herd.

As brilliant and curious beings, goats thrive on mental stimulation and challenges. Providing enrichment activities, such as puzzle feeders, climbing structures, or access to varied vegetation, prevents boredom and taps into their instincts. Mental stimulation is essential to goat welfare, contributing to their overall health and happiness. Understanding the need for cognitive engagement enables herders to implement strategies that enhance the well-being of their goats.

The bond between herder and goat is not one-sided; goats can recognize and respond to human emotions and gestures. Studies have shown that goats are adept at discerning human facial expressions, and they can even differentiate between happy and disgruntled expressions. Establishing trust through positive interactions and consistent care strengthens the human-goat bond, facilitating easier handling and management. Goats, with their perceptive nature, become responsive partners in the shared experience of cohabitation.

While diverse and intricate, goat behavior is not fixed but subject to environmental influences and human interactions. Understanding the malleability of goat behavior allows herders to implement practical training and handling techniques. Positive reinforcement, patience, and consistency are critical elements in training goats, whether for basic commands, milking routines, or veterinary procedures. Establishing a mutual understanding between herder and goat is a testament to the adaptability and intelligence of these remarkable animals.

In conclusion, decoding the language of goats is an art that requires patience, observation, and a deep appreciation for the intricacies of their behavior. From vocalizations to body language, goats communicate a wealth of information about their needs, emotions, and social dynamics. Herders who invest time and effort into understanding and responding to this language forge bonds with their goats that transcend the functional roles of farming. As partners in the journey of coexistence, goats and their herders engage in a silent dialogue that enriches the lives of both, creating a harmonious environment where the language of goats is not just heard but understood and celebrated.

CHAPTER II

Getting Started: Setting Up Your Space

Planning Your Goat Farm

Planning a goat farm is a multifaceted endeavor that requires careful consideration of various factors to ensure the well-being of the herd and the success of the operation. The process begins with a thorough assessment of the goals and objectives of the goat farm, whether it be for meat, milk, fiber, or simply companionship. Understanding the purpose of the farm sets the stage for making informed decisions regarding breed selection, infrastructure, and management practices.

The choice of location is a pivotal aspect of planning a goat farm. The climate, terrain, and available vegetation all play crucial roles in determining the suitability of a site for goat farming. Goats are adaptable creatures, but their performance and overall health are influenced by environmental factors. A farm located in an arid region may require different considerations than one in a temperate climate. Adequate space for grazing, access to clean water, and protection from extreme weather conditions are fundamental requirements that must be addressed during the planning phase.

Infrastructure design is another critical element in the planning process. Goat shelters should provide adequate protection from the elements, ensuring that goats have a comfortable and safe environment. The layout of the farm, including the placement of shelters, feeding areas, and handling facilities, should be designed with efficiency and convenience in mind. Adequate fencing is essential for creating separate enclosures, managing grazing rotations, and preventing unauthorized access, whether from predators or neighboring animals.

The selection of goat breeds is a critical decision influencing the farm's overall success. Different breeds have distinct characteristics and requirements, and the choice should align with the operation's goals. Dairy-focused farms may opt for high-milk-yielding breeds such as Saanen or Nubian, while those emphasizing meat production may choose Boer or Kiko goats. Fiber farms may lean towards Angora or Cashmere goats, each with its unique contribution to the farm's objectives. Understanding the specific needs and traits of the chosen breed allows for tailored management practices that optimize productivity.

Managing the nutritional needs of the herd is a cornerstone of successful goat farming. Developing a balanced diet that caters to the specific requirements of the chosen breed is essential for overall health and productivity. Consideration should be given to the availability of grazing land, supplemental feeding options, and the nutritional content of feeds. Access to minerals and vitamins is crucial, and herders must be vigilant in preventing deficiencies that can adversely affect goat health. Planning for sustainable forage management, including rotational grazing practices, ensures a consistent and nutritious food supply for the herd.

Health and veterinary care constitute an integral part of the farm planning process. Establishing a relationship with a veterinarian and implementing a routine health maintenance program are essential for preventing and addressing potential health issues. Vaccination schedules, parasite control measures, and emergency response plans should be established to safeguard the well-being of the herd. Proper record-keeping of health events and treatments aids in the continuous assessment of the herd's health status, enabling timely interventions and adjustments to management practices.

Reproduction management is crucial for goat farms, especially those aiming to sustain or expand their herd. Planning breeding seasons, managing the mating process, and caring for pregnant and nursing require attention to detail. Understanding the signs of heat, implementing controlled breeding programs, and ensuring appropriate nutrition for breeding females contribute to successful reproduction. Herders should also plan to care for newborn kids, addressing their nutritional needs and providing a conducive environment for their growth and development.

Financial planning is integral to the sustainability of a goat farm. Budgeting for initial setup costs, ongoing operational expenses, and potential emergencies ensures that the farm remains economically viable. Consideration should be given to the costs associated with infrastructure, fencing, feeding, veterinary care, and equipment. Additionally, forecasting potential income streams, whether from the sale of meat, milk, fiber, or breeding stock, aids in making informed financial decisions and setting realistic expectations for the farm's profitability.

Marketing strategies should be considered during the planning phase, especially for farms intending to sell goat products or breeding stock. Building a brand identity, establishing an online presence, and networking within the goat farming community contribute to successful marketing. Understanding local regulations and market demands helps tailor marketing efforts to the target audience. Whether selling directly to consumers, supplying local markets, or participating in goat-related events, effective marketing is essential for reaching the intended audience and maximizing the farm's market potential.

Environmental sustainability is an increasingly important aspect of farm planning. Implementing eco-friendly practices, such as waste management, energy conservation, and sustainable land use, aligns with the broader goals of responsible farming. Utilizing resources efficiently, minimizing environmental impact, and adopting practices that promote biodiversity contribute to the long-term health of the farm and its surrounding ecosystem.

Community engagement and networking play a vital role in the success of a goat farm. Establishing connections with local goat enthusiasts, agricultural organizations, and community groups creates a support network that can provide valuable insights, assistance, and opportunities for collaboration. Participation in events, workshops, and farmer's markets fosters a sense of community and facilitates the exchange of knowledge and experiences among goat herders.

In conclusion, planning a goat farm is a holistic process that involves careful consideration of numerous factors, each playing a crucial role in the overall success of the operation. From defining the farm's purpose and selecting a suitable location to choosing the right breed, designing infrastructure, and implementing sustainable practices, every decision contributes to the farm's viability and productivity. Successful goat farming requires a blend of practical knowledge, thoughtful planning, and a deep appreciation for the unique characteristics and needs of these remarkable animals. By embracing a comprehensive and strategic approach, herders can cultivate a thriving goat farm that not only meets their goals but also contributes to the well-being of the animals and the broader agricultural community.

Building Suitable Shelters

Creating suitable shelters for goats is a critical aspect of responsible and effective goat farming. These structures serve as the primary refuge for goats, offering protection from the elements, ensuring their well-being, and contributing to the overall success of the farm. The design and construction of goat shelters should align with the specific needs and characteristics of the herd, considering factors such as climate, terrain, and the chosen breed. Adequate shelter not only safeguards goats from extreme weather conditions but also provides a comfortable space for rest, feeding, and social interactions.

When planning and building shelters, the first consideration is the region's prevailing climate. Insulated and wind-resistant structures protect goats from harsh winter conditions in colder climates. This may involve incorporating features such as closed sides, insulated roofing, and sufficient bedding to ensure goats stay warm during cold spells. In contrast, in hot and arid climates, shelters should focus on providing shade and good ventilation to prevent heat stress. Raised platforms and open-sided structures allow air circulation, creating a relaxed and comfortable environment for goats during the hotter months.

The size of the shelter is a crucial factor in ensuring the comfort and well-being of the goats. Overcrowded conditions can lead to stress, aggression, and the spread of diseases within the herd. The shelter should provide enough space to accommodate the entire herd comfortably, allowing them to lie down, move around, and access feeding and watering areas without congestion. Additionally, a well-designed shelter layout considers the social dynamics of goats, ensuring that dominant and submissive individuals have separate spaces to reduce conflicts.

The construction materials used for goat shelters play a pivotal role in their durability and functionality. Sturdy and weather-resistant materials, such as treated wood, metal, or durable plastics, are commonly employed. Goats have a penchant for nibbling on surfaces, so materials should also be non-toxic and resistant to damage from chewing. Proper insulation materials can help regulate temperature within the shelter, providing a more comfortable environment for goats in varying weather conditions.

Ventilation is a critical consideration in shelter design, irrespective of climate. Good airflow helps prevent humidity buildup, reduces the risk of respiratory issues, and ensures a fresh and healthy environment. Adequate ventilation can be achieved through strategically placed windows, vents, or open sides, allowing for natural air exchange while protecting from drafts and rain. This becomes especially crucial in regions with high humidity or during inclement weather.

Accessibility to clean water is paramount for goat health, and shelter design should incorporate a convenient and reliable water source. This may involve installing automatic waterers or ensuring that water troughs are easily accessible within the sheltered area. Proper drainage systems are also essential to prevent water pooling around the shelter, minimizing the risk of mud, odors, and potential health issues.

The flooring within the shelter should be carefully chosen to facilitate cleanliness and hygiene. Goats are susceptible to hoof-related issues, and standing on damp or soiled surfaces can lead to foot problems. Well-drained floors, such as those made from gravel or concrete, allow for easy cleaning and reduce the risk of bacterial growth. Bedding materials, such as straw or hay, can be added to provide additional comfort and insulation, particularly during colder months.

Considering the social nature of goats, the shelter design should encourage positive interactions and minimize stress. Individual feeding stations, comfortable resting areas, and spaces for socializing contribute to a harmonious environment within the shelter. Goats prefer elevated areas for relaxing, so providing platforms or raised structures allows them to exhibit natural behaviors and promotes a sense of security.

In instances where goats are raised for specific purposes, such as dairy or meat production, additional considerations come into play. Milking parlors and facilities for storing equipment and feed may need to be incorporated into the shelter design. Similarly, if goats are being raised for meat, the shelter should be conveniently located near processing facilities, reducing stress during transportation.

Regular maintenance and cleanliness are fundamental aspects of goat shelter management. Timely removal of manure and soiled bedding prevents the buildup of odors, reduces the risk of disease, and ensures a hygienic environment for the goats. Routine inspections for signs of wear and tear, such as damaged roofing or fencing, allow for prompt repairs and prevent potential hazards. Regular cleaning of feeding and watering areas helps maintain a sanitary environment, contributing to the overall health and well-being of the herd.

In conclusion, building suitable shelters for goats is a meticulous and strategic process that encompasses a range of factors, from climate considerations to herd dynamics and functionality. Well-designed shelters provide a sanctuary for goats, protecting them from the elements and fostering an environment conducive to their health and natural behaviors. By considering the unique needs of the herd, employing durable and appropriate materials, and incorporating features that promote cleanliness and comfort, herders can create shelters that not only meet the physical needs of the goats but also contribute to the success and sustainability of the overall farming operation.

Creating a Goat-Friendly Environment

Creating a goat-friendly environment is a holistic approach that goes beyond merely providing shelter and sustenance. It involves crafting a space that accommodates the natural behaviors, social dynamics, and health requirements of goats. The success of a goat farm is intricately tied to the overall well-being of the herd, and a thoughtful environment fosters contentment, minimizes stress, and contributes to the productivity and longevity of the goats.

Land Management: The foundation of a goat-friendly environment begins with thoughtful land management. Goats are avid foragers, and ample space for grazing and browsing is essential. Rotational grazing practices prevent overgrazing, allowing vegetation to regenerate and ensuring a constant and nutritious food supply. Access to diverse forage types contributes to a well-balanced diet, supporting the herd's nutritional needs.

Fencing: Well-designed fencing is critical to creating a secure and goat-friendly environment. Goats are known for their curiosity and knack for exploration, making sturdy fencing crucial for preventing escapes and protecting them from potential hazards. The fence's height, durability, and design should be tailored to the specific breed and size of the goats, ensuring their safety and minimizing the risk of injury.

Enrichment: Goats thrive on mental stimulation and opportunities for play. Introducing enrichment activities, such as climbing structures, platforms, and toys, engages their instincts and prevents boredom. These activities contribute to the goats' mental well-being and promote physical exercise, reducing the risk of obesity and related health issues.

Shelter and Rest Areas: Providing comfortable and strategically located shelters is essential for creating a goat-friendly environment. Shelters should offer protection from adverse weather conditions while allowing for proper ventilation. Additionally, goats prefer elevated surfaces for resting, so incorporating platforms or raised areas within the shelter will enable them to exhibit their natural behaviors.

Social Dynamics: Goats are highly social animals, and the environment's design should accommodate their need for companionship. Overcrowding should be avoided to prevent stress and aggression within the herd. Creating separate spaces for dominant and submissive individuals creates a more harmonious social structure. Paying attention to the dynamics between goats and introducing them to new herd members gradually fosters positive relationships.

Water Accessibility: Adequate access to clean water is fundamental for the health and well-being of goats. Placing water troughs in multiple locations ensures that all members of the herd can hydrate comfortably. Regular cleaning and monitoring of water sources prevent contamination and contribute to overall herd health.

Mineral Stations: Goats have specific nutritional needs, and providing access to mineral stations is vital for meeting these requirements. Mineral blocks or loose minerals tailored to the dietary needs of goats should be readily available. Regular monitoring of mineral consumption helps identify and address any deficiencies.

Manure Management: Efficient management is crucial for maintaining a clean and hygienic environment. Regular removal of manure prevents the buildup of odors, reduces the risk of parasite infestations, and minimizes environmental impact. Proper disposal or composting of manure contributes to sustainable land management practices.

Veterinary Care: Creating a goat-friendly environment extends to proactive veterinary care. Establishing a relationship with a veterinarian and implementing routine health checks contribute to the overall health and longevity of the herd. Timely vaccinations, parasite control measures, and prompt attention to signs of illness or distress are essential components of a comprehensive veterinary care plan.

Kid-Friendly Areas: When raising kids, creating specific areas tailored to their needs is crucial. Providing secure spaces where kids can play, explore, and rest without interference from adult goats ensures their safety and development. Additionally, maintaining a separate feeding area for kids guarantees they receive the appropriate nutrition for their growth.

Environmental Considerations: Sustainable and eco-friendly practices should be integral to creating a goat-friendly environment. This involves managing vegetation, preventing soil erosion, and implementing practices that promote biodiversity. By considering the broader environmental impact of goat farming, herders contribute to the health of the surrounding ecosystem.

Education and Observation: Creating a goat-friendly environment requires continuous education and observation. Herders should stay informed about the specific needs of their chosen breed, be updated on advancements in goat farming, and actively observe the behaviors and interactions within the herd. This ongoing engagement allows herders to adapt their management practices, address emerging issues, and continually refine the environment for the well-being of the goats.

In conclusion, creating a goat-friendly environment is a comprehensive and dynamic process that involves a harmonious integration of various elements. From land management and fencing to enrichment activities, shelter design, and veterinary care, every aspect contributes to the overall well-being of the herd. By recognizing and accommodating the natural instincts, social dynamics, and health requirements of goats, herders not only enhance the quality of life for their animals but also lay the foundation for a successful and sustainable goat farming operation. A goat-friendly environment is more than a physical space; it is a testament to the thoughtful and conscientious stewardship that defines responsible goat husbandry.

CHAPTER III

Selecting Your Herd

Choosing the Right Breeds for Your Goals

Selecting the suitable breeds is a pivotal decision in the journey of goat farming, as it shapes the trajectory of the entire operation. The diverse world of goats offers a multitude of breeds, each with distinct characteristics, purposes, and adaptations. Understanding your goals and the traits required for your intended outcomes is fundamental in making informed decisions about the breeds you choose to raise. High milk-yielding breeds such as Saanen, Nubian, or Alpine may be the preferred choice for those venturing into dairy farming. These breeds are renowned for their prolific milk production, making them valuable assets for dairy enthusiasts seeking to establish a thriving milking herd. On the other hand, if meat production is the primary objective, the robust Boer goat from South Africa stands out for its impressive growth rates and high-quality meat. Boers have gained global popularity for their adaptability and efficiency in converting forage into muscle, making them a preferred choice for meat-focused operations.

Angora and Cashmere goats take center stage for those drawn to the allure of luxurious fiber production. Angoras, with their long, silky mohair, are ideal for fiber enthusiasts seeking to embark on the artistry of spinning and weaving. Their elegant appearance and gentle temperament make them valuable additions to farms focusing on specialty fibers. Cashmere goats from the Himalayan region contribute to the textile industry with their fine undercoat, prized for its softness and warmth. Understanding the specific fiber qualities and grooming requirements of each breed is crucial for those aspiring to venture into the world of fiber production.

Beyond the primary purposes of meat, milk, and fiber, some breeds serve multifaceted roles, catering to the diverse needs of farmers. Nigerian Dwarf goats, despite their diminutive size, are renowned for their high milk production relative to their body size. This makes them an excellent choice for small-scale dairy operations or those with limited space. Their friendly disposition and adaptability enhance their appeal, especially for farmers seeking goats as pets or companion animals.

Environmental considerations also play a role in breed selection. Goats are remarkably adaptable animals, and different breeds thrive in varying climates and terrains. Alpine goats, originating in the Swiss Alps, exhibit resilience in mountainous regions and are well-suited for challenging terrains. Spanish goats, descendants of those brought by Spanish explorers to North America, are known for their hardiness and ability to thrive in arid conditions, making them valuable assets for sustainable land management.

Understanding the behavioral traits of different breeds is equally crucial. Goats exhibit diverse personalities; certain breeds may be more docile, while others may display more assertive or independent behaviors. LaMancha goats, recognized for their distinct ear characteristics, are known for their calm and gentle demeanor, making them suitable for various farming environments. However, the spirited nature of breeds like the Kiko goat, developed in New Zealand for meat production, may be better suited for farmers seeking more independent and hardy animals.

When selecting breeds, it is essential to consider the size of the operation and the available resources. Larger farms with extensive grazing land may successfully manage larger breeds such as Toggenburg or Boer goats. However, smaller operations or urban farms may lean towards smaller breeds like Pygmy or Nigerian Dwarf goats, which require less space and are often well-suited for backyard farming. The available resources for feed, healthcare, and infrastructure should align with the specific needs and characteristics of the chosen breeds to ensure their optimal health and productivity.

Additionally, breeding goals should be carefully considered. Some breeds are recognized for specific genetic traits, and herders may opt for selective breeding to enhance desirable characteristics in their herd. For instance, a farmer aiming to develop a hardy and adaptable herd may selectively breed from Spanish or Kiko goats, emphasizing traits such as resistance to parasites and efficient foraging abilities.

Market demand and regional preferences also influence breed selection. Understanding the market for goat products in a particular region, whether for meat, milk, or fiber, guides farmers in choosing breeds that align with consumer preferences. Participating in local markets, understanding consumer trends, and networking within the goat farming community can provide valuable insights into the demand for specific breeds and products.

Ultimately, the choice of goat breeds is a dynamic and strategic decision that requires a comprehensive understanding of the farm's goals, environmental conditions, and the specific traits of each breed. While breed selection sets the foundation for the farm's success, ongoing management practices, including nutrition, healthcare, and environmental considerations, are integral in realizing the full potential of the chosen breeds. The diversity within the world of goats offers a rich palette for farmers to sculpt their operations, whether pursuing the elegance of fiber, the abundance of milk, the juiciness of meat, or the joy of companionship. By aligning breed selection with the farm's goals and fostering a deep understanding of the unique characteristics of each breed, farmers pave the way for a flourishing and purposeful goat farming venture.

Evaluating Healthy Goats: Tips for Selection

Evaluating goats' health is a fundamental skill that every herder should possess, as the herd's well-being is paramount for a successful and sustainable goat farming operation. When selecting goats, whether for breeding, dairy, meat, or as companions, a keen eye for physical indicators, combined with an understanding of behavioral cues, contributes to identifying healthy and thriving individuals. The goat's overall body condition is one of the first aspects to assess. A healthy goat should exhibit a well-proportioned body with a sleek coat and good muscling. Emaciation or excessive thinness may indicate malnutrition or health issues, while obesity can lead to various health complications. The coat should be shiny and free of mats or clumps, and the skin should be elastic, reflecting proper hydration. Lice, ticks, or other external parasites can indicate suboptimal living conditions or neglect.

Observing the eyes and mucous membranes provides insights into the goat's health status. Bright, alert eyes with clear corneas are signs of good health, while dull or cloudy eyes may indicate underlying issues. The color of the mucous membranes, particularly the gums, should be a healthy pink, indicating proper oxygenation. Paleness or jaundice may signal anemia or liver problems. The capillary refill time, measured by pressing on the gums and observing how quickly the color returns, should be rapid, indicating good blood circulation.

Assessing respiratory health involves observing the goat's breathing patterns. Average respiratory rates for goats typically range from 12 to 20 breaths per minute. Labored breathing, coughing, or nasal discharge may be signs of respiratory infections or other respiratory issues. Additionally, examining the goat's ears for signs of discharge or swelling and checking for unusual odors can further indicate health or potential problems.

Checking the hooves is a crucial aspect of evaluating a goat's health. Overgrown or misshapen hooves can lead to discomfort, lameness, and various health problems. Regular hoof trimming is essential to prevent hoof rot or abscesses. Additionally, inspecting the udder in female goats is imperative for those involved in dairy farming. A healthy udder should be well-formed, symmetrical, and free from lumps, swelling, or signs of infection. Any abnormalities may affect milk production and quality.

Behavioral observations are equally crucial in gauging a goat's health. A healthy goat should exhibit alertness, curiosity, and an interest in its surroundings. Lethargy, isolation from the herd, or a sudden behavior change may indicate underlying health issues. Healthy goats should also display a regular eating pattern, with a consistent appetite and the ability to chew cud. A sudden loss of appetite or changes in eating habits can signal potential health problems such as dental issues, digestive disorders, or infections.

When selecting breeding goats, reproductive health becomes a critical consideration. Evaluating the reproductive organs, particularly in does, ensures that they can successfully breed. Assessing the testicles in bucks is equally essential, as any abnormalities or signs of injury may affect fertility. Understanding the reproductive history of a goat, including previous pregnancies and kidding experiences, can provide insights into its reproductive health and potential complications.

Regular veterinary check-ups and health records are valuable tools for evaluating a goat's health. A thorough examination by a veterinarian can identify underlying health issues, assess vaccination status, and guide preventive care measures. Health records, including vaccination history, deworming schedules, and any past illnesses or treatments, offer a comprehensive view of a goat's health and enable herders to make informed decisions about their selection and management.

It's essential to consider the age and life stage of the goat when evaluating its health. Kids should exhibit lively behavior, with proper weight gain and development. Older goats may show signs of aging, including dental wear, decreased muscle mass, or joint stiffness. Understanding the specific needs and considerations for each life stage enables herders to tailor their care and management practices to support the health and well-being of their goats throughout their lifecycles.

Finally, quarantine and biosecurity measures are crucial when introducing new goats to a herd. Isolating new arrivals for a period allows for observing and detecting any potential health issues before they can spread to the existing herd. Implementing biosecurity practices, such as regular handwashing, disinfecting equipment, and controlling access to the farm, helps prevent the introduction and spread of diseases.

In conclusion, evaluating healthy goats is a multifaceted process that requires a combination of keen observation, physical examination, and an understanding of behavioral cues. A holistic approach that considers the immediate physical condition and the overall well-being, reproductive health, and life stage of the goat is essential. Regular veterinary care, health records, and biosecurity measures further contribute to successfully evaluating and managing goat health. By honing the skill of assessing and selecting healthy goats, herders lay the foundation for a resilient and thriving herd that can contribute to their goat farming venture's overall success and sustainability.

Building Your Ideal Herd Composition

Building the ideal herd composition is a strategic endeavor that involves thoughtful planning, consideration of specific goals, and a deep understanding of the characteristics and dynamics of different goat breeds. The composition of a goat herd can vary widely based on the objectives of the farm, whether it's focused on meat production, milk, fiber, or a combination of these. Selecting high-yielding dairy goat breeds such as Saanen,

Nubian, or Alpine is paramount for a dairy-oriented operation. These breeds are renowned for their milk production, producing rich and flavorful milk that can be used for various dairy products. In contrast, a meat- centric herd may prioritize breeds like Boer or Kiko goats, known for their robust growth rates and high-quality meat. Boer goats, in particular, have gained international acclaim for efficiently converting forage into muscle, making them a preferred choice for meat production.

The size of the herd also plays a crucial role in its composition. Smaller farms or those with limited space may opt for smaller breeds like Nigerian Dwarf or Pygmy goats, which require less space and are also known for their amicable nature, making them suitable for smaller-scale or hobbyist farms. On the other hand, more extensive operations may succeed in managing larger breeds, such as Toggenburg or Alpine goats, which can contribute to higher milk production and meat yields.

Considering the environmental conditions of the farm is vital in determining the ideal herd composition. Certain breeds are better adapted to specific climates and terrains. Alpine goats, with their origins in the Swiss Alps, exhibit resilience in mountainous regions, while Spanish goats have proven their hardiness in arid conditions. Selecting breeds that thrive in the local climate ensures the overall health and productivity of the herd, minimizing the need for extensive management interventions.

Reproductive considerations play a significant role in building the ideal herd composition. Understanding the reproductive capabilities and characteristics of different breeds aids in planning breeding programs that align with the farm's goals. Bucks from breeds known for their reproductive efficiency may be introduced to the herd to enhance breeding outcomes. For farms that maintain a self-sustaining herd, selecting those with strong maternal instincts and successful kidding histories contributes to a thriving and expanding population.

The diversity within the goat world allows for the creation of multifunctional herds that cater to various needs simultaneously. A farm may incorporate dual-purpose breeds that excel in milk and meat production. For example, the Oberhasli goat, recognized for its striking appearance and medium-sized stature, is proficient in milk and meat production. Dual-purpose breeds offer flexibility and versatility, allowing herders to optimize resources and meet diverse market demands.

Behavioral traits and social dynamics also play a crucial role in building an ideal herd composition. Some breeds exhibit more docile and calm behavior, making them suitable for smaller farms or those with family-oriented setups. LaMancha goats' distinctive short ears are known for their gentle temperament, making them popular choices for farms where ease of handling is a priority. Understanding a herd's social hierarchy and dynamics is essential to prevent conflicts and ensure harmonious interactions among the goats.

Introducing variety within the herd composition can contribute to a visually appealing and exciting farm environment. Different coat colors, patterns, and markings add aesthetic value and can be a unique selling point for farms engaged in agritourism or those supplying goats for petting zoos. Additionally, diverse coat colors may have cultural or symbolic significance, adding a layer of richness to the overall farming experience.

Building an ideal herd composition also involves careful consideration of the specific needs and goals of the herders. Goat farming can be a deeply personal and fulfilling endeavor, and the composition of the herd should align with the herder's aspirations and preferences. Whether it's the elegance of Angora goats contributing to the fiber arts or the companionship offered by Nigerian Dwarf goats, the herder's individual goals and inclinations shape the ideal herd's composition.

The concept of sustainable and ethical farming practices is increasingly influencing herd composition. Farms that prioritize ethical considerations may opt for breeds well-suited to pasture-based systems and emphasize rotational grazing. Implementing regenerative agricultural practices, such as rotational grazing, contributes to the health of the land and the well-being of the goats. Breeds with strong foraging abilities, such as Kiko goats, are particularly suited for such systems, thriving on a diverse vegetation diet.

Adaptability and resilience are critical considerations in the face of environmental challenges and changing conditions. Herd composition should reflect the ability of the goats to adapt to fluctuations in climate, feed availability, and disease resistance. Incorporating breeds known for their hardiness and adaptability ensures a robust and resilient herd that can withstand unforeseen challenges.

In conclusion, building the ideal herd composition is a nuanced and dynamic process that requires a combination of strategic planning, knowledge of breed characteristics, and alignment with the goals and values of the herd. Whether focused on meat, milk, fiber, or a combination of these, the selection of breeds, the size of the herd, and consideration of environmental and ethical factors all contribute to the composition of a thriving and purposeful goat herd. By understanding the unique traits of each breed and tailoring the herd to the specific needs and objectives of the farm, herders can cultivate a diverse and resilient herd that meets their production goals and aligns with their vision of responsible and sustainable goat farming.

CHAPTER IV

Nutrition Essentials

Crafting a Balanced Goat Diet

Crafting a balanced diet for goats is a fundamental aspect of responsible goat husbandry, influencing their overall health, productivity, and well-being. Goats are known for their diverse dietary preferences, often displaying selective feeding habits. As ruminants, their digestive systems are uniquely adapted to process fibrous plant material efficiently, making forage the cornerstone of their diet. High-quality forage, such as pasture grasses, legumes, and browse, provides essential nutrients, including fiber, vitamins, and minerals. Pasture management practices, such as rotational grazing, contribute to sustained forage availability and prevent overgrazing, ensuring a continuous and nutritious food supply.

In addition to forage, goats require a balanced concentrate feed to meet their nutritional needs, especially when specific production goals, such as milk or meat production, are targeted. The composition of concentrate feeds varies based on the purpose of the herd. Dairy goats, for instance, benefit from concentrates rich in protein and energy to support milk production. Commercial goat feeds formulated for different life stages, such as kid, growing, and lactating formulations, provide a convenient and nutritionally balanced option. Careful consideration should be given to the protein content, energy levels, and essential minerals in concentrate feeds to ensure they align with the nutritional requirements of the goats.

Supplemental feeding may be necessary, particularly in regions where forage quality is variable or during periods of nutritional stress, such as winter. Providing mineral supplements, including salt blocks fortified with essential minerals, prevents deficiencies that can adversely affect goat health. Selenium, copper, zinc, and other trace minerals are vital for overall well-being and productivity. However, over-supplementation should be avoided, as excessive intake of certain minerals can lead to toxicity.

Water is a non-negotiable component of a balanced goat diet. Adequate access to clean and fresh water is essential for digestion, nutrient absorption, and thermoregulation. Goat water requirements vary based on age, size, reproductive status, and environmental conditions. Maintaining clean water sources, especially during extreme weather conditions, prevents dehydration and supports overall goat health.

The nutritional needs of goats also depend on their life stage. Kids, for example, have specific dietary requirements for growth and development. Adequate colostrum intake in the first hours of life provides essential antibodies and nutrients. As kids transition to solid food, a diet rich in protein and energy supports their rapid growth. Lactating does have increased nutritional demands, requiring higher levels of energy, protein, and minerals to support milk production. Balanced diets during gestation contribute to the health of the developing fetus and set the stage for successful kidding.

Understanding the specific nutritional needs of different breeds is integral to crafting a balanced diet. With their higher milk production, dairy goats demand diets with increased protein and energy content. While crucial for overall digestive health, fiber should not compromise nutrient density. Meat goats, on the other hand, benefit from diets that support efficient muscle development and growth. Fiber remains essential for maintaining gut health, and the energy and protein content of the diet should align with meat production goals.

For goats involved in fiber production, such as Angora or Cashmere goats, the diet should prioritize quality forage to support their unique nutritional needs. Adequate protein levels contribute to healthy fiber growth, and supplementary feeds may be necessary to ensure proper nutrition. Fiber-producing goats also require meticulous grooming to prevent matting and maintain the quality of their fleece.

Factors such as season, climate, and available forage can influence goats' feeding behavior. During periods of drought or winter, when forage availability may be limited, supplementary feeding becomes crucial to prevent nutrient deficiencies. Adjusting the diet based on seasonal variations ensures that goats receive the necessary nutrients year-round.

Implementing a feeding management plan that aligns with the specific goals of the herd is essential for long-term success. Monitoring body condition, weight gain or loss, and reproductive performance aids in fine-tuning the feeding program. Regular veterinary consultations and nutritional assessments provide valuable insights into the health and nutritional status of the herd, guiding adjustments to the feeding regimen.

In conclusion, crafting a balanced diet for goats involves a dynamic and multifaceted approach, considering their unique dietary preferences, nutritional needs, and production goals. Each component is crucial in supporting overall goat health and productivity, from high-quality forage to well-formulated concentrate feeds and mineral supplements. A balanced diet not only contributes to the prevention of nutritional deficiencies but also enhances the resilience of the herd, allowing goats to thrive in diverse environments and conditions. By combining a deep understanding of goat nutrition with effective feeding management practices, herders can cultivate a herd that meets their production objectives and exemplifies the principles of responsible and sustainable goat farming.

Understanding Nutritional Requirements

Understanding the nutritional requirements of goats is foundational to successful and responsible goat husbandry. As ruminant animals, goats possess a unique digestive system that efficiently extracts nutrients from fibrous plant material. Most of their diet comes from forage, including pasture grasses, legumes, and browse. Forages provide essential nutrients such as fiber, vitamins, and minerals, contributing to overall digestive health and well-being. Pasture management practices, like rotational grazing, ensure a continuous and diverse forage supply and promote sustainable land use by preventing overgrazing.

In addition to forage, goats require concentrated feeds to meet their specific nutritional needs, especially in scenarios where production goals, such as milk or meat production, are targeted. The composition of these concentrates varies based on the purpose of the herd. For instance, dairy goats benefit from concentrates rich in protein and energy to support the demands of milk production. Commercial goat feeds formulated for different life stages, including kid, growing, and lactating formulations, offer a convenient and nutritionally balanced option. It is crucial to consider protein content, energy levels, and essential mineral composition in concentrates to ensure they align with the unique nutritional requirements of goats.

Mineral supplements play a pivotal role in meeting goats' nutritional needs. Essential minerals, including selenium, copper, zinc, and others, are vital for various physiological functions, such as immunity, reproduction, and bone development. Goats often require supplementation, as the levels of these minerals in forages may be insufficient to meet their needs. However, caution must be exercised to prevent over-supplementation, as excessive intake of certain minerals can lead to toxicity and adverse health effects.

Water is a fundamental component of a goat's diet and is critical for digestion, nutrient absorption, and thermoregulation. Adequate access to clean and fresh water is essential for preventing dehydration and supporting overall health. Goat water requirements vary based on age, size, reproductive status, and environmental conditions. Regular monitoring of water sources and ensuring cleanliness contribute to the overall well-being of the herd.

The nutritional needs of goats are intricately tied to their life stage. Kids, in their early days, rely on colostrum, a nutrient-rich substance produced by the doe shortly after kidding. Colostrum provides essential antibodies and nutrients, supporting the newborn's immune system and overall health. As kids transition to solid food, their diet must be rich in protein and energy to support growth and development.

Lactating does have increased nutritional demands to support milk production. Diets for lactating should be formulated to provide higher levels of energy, protein, and essential minerals. Proper nutrition during lactation supports milk production and contributes to the health and vitality of both the doe and the nursing kids.

It also has specific nutritional requirements during gestation to support the developing fetus. Adequate protein, energy, and mineral levels are crucial during this period to ensure proper fetal growth and development. Good nutrition during gestation positively influences the doe's and her offspring's health and productivity.

Meat goats, bred for efficient growth and muscle development, require diets that support these goals. Protein and energy content become crucial factors in the formulation of diets for meat-producing goats. Well-balanced nutrition contributes to optimal growth rates and the overall quality of the meat produced.

Fiber-producing goats like Angora or Cashmere goats have unique nutritional needs for fiber production. Diets for these goats must be designed to support the health of their fleece. Adequate protein levels contribute to healthy fiber growth, and supplementary feeds may be necessary to ensure proper nutrition for optimal fiber quality.

The understanding of specific nutritional needs extends to considerations of different breeds. Dairy goats, for example, with their higher milk production, demand diets with increased protein and energy content. Fiber remains essential for overall digestive health, but nutrient density should not be compromised. In contrast, meat goats prioritize diets that support efficient muscle development and growth. Balancing the energy and protein content of the diet aligns with meat production goals.

Environmental factors influence goats' nutritional requirements: seasonal variations, climate conditions, and forage availability impact dietary needs. During drought or winter, when forage quality may be compromised, supplementary feeding becomes crucial to prevent nutrient deficiencies. Adjusting diets based on these seasonal variations ensures that goats receive the necessary nutrients year-round.

Reproductive considerations are integral to understanding nutritional requirements. Ensuring proper nutrition for breeding bucks contributes to their reproductive efficiency. Monitoring body condition, weight, and overall health aids in optimizing breeding outcomes. Understanding the reproductive cycle of does and tailoring their diets to support successful gestation, kidding, and lactation is essential for sustaining a healthy and productive herd.

Regular veterinary consultations, nutritional assessments, and monitoring of body condition scores contribute to a comprehensive understanding of goats' nutritional status. The feeding program can be adjusted based on these assessments to ensure that the goats receive optimal nutrition for their specific needs.

In conclusion, understanding the nutritional requirements of goats is a dynamic and multifaceted endeavor that involves a combination of forage management, balanced concentrates, mineral supplementation, and attentive care to specific life stages and production goals. The nutritional foundation for goats directly influences their health, productivity, and overall well-being. By tailoring diets to meet the unique needs of different breeds, life stages, and environmental conditions, herders can foster a herd that thrives and exemplifies the principles of responsible and sustainable goat farming.

Common Feeding Mistakes to Avoid

Avoiding common feeding mistakes is paramount for maintaining a goat herd's health, productivity, and overall well-being. One prevalent error is inadequate access to high-quality forage, which is the foundation of a goat's diet. Bad pasture management, including overgrazing or limited access to fresh forage, can result in nutritional deficiencies and compromised digestive health. Proper rotational grazing practices, coupled with a diverse mix of pasture grasses, legumes, and browse, ensure a continuous and nutritious forage supply, meeting the goats' essential nutritional needs.

Another critical mistake is providing unbalanced or inappropriate concentrates. Concentrate feeds that do not align with the specific nutritional requirements of goats, especially during different life stages or production goals, can lead to deficiencies or excesses in essential nutrients. For example, feeding dairy goats a concentrate formulated for meat goats may result in inadequate protein and energy levels for optimal milk production. To avoid this, herders should carefully select commercial goat feeds tailored to the specific needs of their herd, adjusting formulations based on the goals and life stages of the goats.

Mineral deficiencies or imbalances are common pitfalls in goat nutrition. Goats require essential minerals, such as selenium, copper, and zinc, for various physiological functions. Inadequate access to these minerals in forages or failure to provide appropriate mineral supplements can result in health issues, including impaired immune function, reproductive problems, and skeletal disorders. Regularly monitoring and addressing mineral deficiencies through supplementation is crucial for maintaining the overall health and well-being of the herd.

Water, a fundamental component of a goat's diet, must be addressed, leading to dehydration and related health problems. Inconsistent access to clean and fresh water can compromise digestion, nutrient absorption, and thermoregulation. Herders should ensure that water sources are regularly checked and cleaned, especially during extreme weather conditions, to prevent dehydration and support overall goat health.

Overfeeding or underfeeding goats is a common mistake that can have significant consequences. Overfeeding can lead to obesity, which is associated with various health issues, including metabolic disorders and reduced reproductive efficiency. On the other hand, underfeeding can result in malnutrition, stunted growth, and decreased productivity. Monitoring body condition scores, adjusting feeding regimens based on individual needs, and considering the energy requirements of different life stages are crucial aspects of preventing overfeeding or underfeeding.

Neglecting the nutritional needs of pregnant and lactating women is a prevalent mistake with serious repercussions. Proper nutrition during gestation is essential for fetal development, and inadequate feeding can result in weak kids, increased kidding difficulties, or even losses. Lactating does have raised energy and nutrient requirements to support milk production. Failure to meet these demands can lead to poor milk quality, decreased kid growth rates, and compromised doe health. Adjusting diets to meet the specific needs of breeding is vital for ensuring successful pregnancies and healthy offspring.

Another common feeding mistake is relying solely on visual assessments without regular veterinary check-ups. While visual indicators such as body condition scores provide valuable insights, they may not capture underlying health issues or nutrient deficiencies. Regular veterinary consultations, nutritional assessments, and diagnostic testing contribute to a more comprehensive understanding of the herd's nutritional status. Herders should work collaboratively with veterinarians to monitor their goats' health and dietary needs and make informed adjustments to their feeding programs.

More attention to pasture management can lead to nutritional imbalances and health issues. Overgrazing and poor pasture quality can result in nutrient deficiencies, while excessive reliance on certain forages may lead to imbalances in the goat's diet. Implementing sustainable pasture management practices, including rotational grazing, maintaining diverse forage varieties, and addressing soil fertility, is crucial for preventing common feeding mistakes related to pasture quality and availability.

It is essential to account for environmental factors, such as climate and seasonal variations, to ensure the nutritional well-being of goats. Extreme weather conditions, such as drought or winter, can affect forage quality and availability, necessitating adjustments to feeding regimens. Herders should proactively adapt their feeding programs based on environmental factors to ensure goats receive adequate nutrition year-round.

Underestimating the impact of stress on goat nutrition is a mistake that can compromise herd health. Changes in herd dynamics, transportation, or environmental disturbances can induce stress, affecting feed intake and nutrient utilization. Mitigating stress through proper management practices, gradually introducing new feed sources, and providing a stable and comfortable environment contribute to consistent feed consumption and overall well-being.

In conclusion, avoiding common feeding mistakes is crucial for a goat herd's health, productivity, and longevity. Herders should prioritize proper forage management, select balanced concentrates, address mineral deficiencies, ensure consistent access to clean water, and monitor the nutritional needs of different life stages. Regular veterinary consultations, pasture management practices, and adaptability to environmental factors contribute to a comprehensive and responsible approach to goat nutrition. By steering clear of these common feeding pitfalls, herders can foster a resilient and thriving herd that exemplifies the principles of sustainable and responsible goat farming.

CHAPTER V

Health and Veterinary Care

Routine Health Checks: What to Look For

Routine health checks are a cornerstone of responsible goat husbandry, allowing herders to monitor the well-being of their herd and promptly address any potential health issues. Observing goats' general demeanor and behavior provides valuable insights into their health. Healthy goats typically exhibit alertness, curiosity, and an interest in their surroundings. Lethargy, depression, or changes in behavior can be early indicators of underlying health problems. Regular visual assessments also include monitoring body condition scores, ensuring that goats maintain an optimal weight and musculature, which indicates proper nutrition and overall health.

A thorough examination of the eyes and mucous membranes contributes to health assessments. Bright, clear eyes are signs of good health, while dull or cloudy eyes may indicate systemic issues. The color of the mucous membranes, particularly the gums, is an essential diagnostic tool. A healthy goat's gums should be pink, indicating proper oxygenation. Paleness or jaundice may signal anemia or liver problems. The capillary refill time, measured by pressing on the gums and observing how quickly the color returns, should be rapid, reflecting good blood circulation.

Respiratory health is another crucial aspect of routine health checks. Monitoring the breathing patterns of goats helps identify respiratory issues such as infections or allergies. Average respiratory rates for goats typically range from 12 to 20 breaths per minute. Labored breathing, coughing, nasal discharge, or wheezing may indicate respiratory distress. Regular assessment of the ears for signs of discharge, swelling, or unusual odors is also essential, as ear issues can be symptomatic of infections or mites.

Hoof health is integral to overall goat well-being. Regular inspection of hooves helps identify overgrowth, deformed hooves, or signs of hoof rot. Proper hoof trimming, a routine part of goat management, prevents lameness and other hoof-related problems. In female goats, assessing the udder is crucial for those involved in dairy farming. A healthy udder should be well-formed, symmetrical, and free from lumps, swelling, or signs of infection. Any abnormalities may affect milk production and quality.

Behavioral observations during routine health checks include monitoring eating patterns. A healthy goat should exhibit a consistent appetite, chew cud regularly, and have regular bowel movements. Sudden changes in eating habits or signs of gastrointestinal distress may indicate dental issues, digestive disorders, or other health problems. Additionally, observing the social interactions within the herd helps identify any bullying or aggression and signs of goats isolating themselves, which may signal underlying health issues.

Reproductive health assessments are crucial, especially for breeding goats. Evaluating the reproductive organs of does ensures their capability for successful breeding. Assessing the testicles in bucks is equally essential, as any abnormalities or signs of injury may affect fertility. Understanding the reproductive history of goats, including previous pregnancies and kidding experiences, provides insights into their reproductive health and potential complications.

Regular veterinary check-ups complement routine health checks, enabling more in-depth assessments and preventative care measures. Working closely with a veterinarian allows for implementing vaccination schedules, deworming programs, and other preventive health measures tailored to the herd's needs. Veterinary consultations are also opportune times to discuss nutritional requirements, breeding plans, and emerging health concerns.

Health records are indispensable in routine health checks, providing a comprehensive history of each goat's health and management. Keeping detailed records of vaccinations, deworming schedules, reproductive history, and any illnesses or treatments facilitates proactive health management. These records aid in tracking individual and herd health trends, contributing to informed decision-making regarding breeding programs, nutritional adjustments, and overall herd management.

Monitoring for signs of parasitic infestations is essential during routine health checks. Internal parasites, such as gastrointestinal worms, can adversely affect goat health and productivity. Symptoms may include weight loss, poor coat condition, anemia, and diarrhea. Fecal testing and strategic deworming, guided by veterinary advice, help manage and prevent parasite infestations. External parasites, such as lice and mites, should also be monitored, and appropriate measures, including topical treatments, should be implemented if necessary.

Dental health assessments are integral to routine health checks, as dental issues can impact a goat's ability to eat and digest food properly. Regularly examining the teeth for signs of overgrowth, uneven wear, or dental abnormalities helps prevent dental problems. Adjustments to the diet, including providing adequate fiber and offering appropriate dental chews or blocks, contribute to maintaining good dental health.

Foot rot is a common concern in goats, particularly in moist environments. Routine health checks should include inspections for signs of foot rot, such as lameness, swelling, and foul odor. Prompt treatment, often involving foot trimming, cleaning, and topical therapies, helps manage and prevent the spread of foot rot within the herd.

In conclusion, routine health checks are an integral part of responsible goat management, allowing herders to monitor the well-being of their herd and address potential health issues. Regular visual assessments, including observations of behavior, eyes, mucous membranes, respiratory health, and hooves, provide valuable insights into the overall health of goats. Reproductive health assessments, dental checks, and monitoring for parasitic infestations contribute to comprehensive health management. Collaborating with veterinarians, maintaining detailed health records, and implementing preventative care measures further enhance the effectiveness of routine health checks. By prioritizing these assessments, herders can ensure their goat herd's health, productivity, and longevity, fostering a thriving and resilient community of animals.

Vaccination Schedules and Preventive Care

Establishing and adhering to a robust vaccination schedule is a fundamental component of preventive care in goat farming. Vaccinations play a crucial role in safeguarding the health and well-being of goats by preventing or mitigating the impact of various infectious diseases. The specific vaccines and vaccination schedule may vary based on geographic location, herd size, and the particular health challenges in the region. A comprehensive approach to preventive care involves collaboration with a veterinarian to tailor vaccination programs to the unique needs of the herd.

Core vaccinations, considered essential for all goats, typically include protection against diseases like Clostridium perfringens types C and D (enterotoxemia), Clostridium tetani (tetanus), and caseous lymphadenitis (CLA). Enterotoxemia, commonly known as overeating disease, is caused by toxins produced by the bacterium Clostridium perfringens and can result in sudden death in goats. Tetanus, caused by the bacterium Clostridium tetani, is a potentially fatal disease characterized by muscle stiffness and spasms. CLA, caused by

Corynebacterium pseudotuberculosis, leads to abscess formation in lymph nodes and other tissues.

In addition to core vaccinations, region-specific considerations may include vaccines for contagious caprine pleuropneumonia (CCPP), caprine arthritis encephalitis (CAE), and other prevalent infections. CCPP, caused by Mycoplasma capricolum subspecies capripneumoniae, affects the respiratory system and can lead to severe pneumonia in goats. CAE is a viral infection resulting in chronic arthritis, mastitis, and progressive weakness.

The timing of vaccinations is critical for their effectiveness, with initial doses often administered during the early months of a goat's life. Booster shots, given at regular intervals, help maintain immunity levels and reinforce protection against diseases. The specific schedule may vary based on the type of vaccine, the region's disease prevalence, and the goat's age and reproductive status.

Preventive care extends beyond vaccinations to include strategic deworming programs. Internal parasites, such as gastrointestinal worms, significantly threaten goat health and productivity. Regular fecal testing, guided by veterinary advice, helps identify the presence of parasites and determine appropriate deworming protocols. Implementing rotational grazing, maintaining clean and dry living conditions, and avoiding overstocking are additional measures to reduce the risk of parasitic infestations.

Proactive measures to prevent external parasites, such as lice and mites, contribute to overall herd health. Topical treatments, including pour-on formulations or dusting powders, can help manage and prevent infestations. Adequate shelter, proper sanitation, and routine grooming practices are complementary strategies to minimize the risk of external parasites affecting the herd.

Routine hoof care is essential to preventive care, contributing to overall goat well-being. Regular hoof trimming helps prevent overgrowth, deformed hooves, and hoof rot. Proper drainage in living areas and providing elevated resting spaces minimize the risk of moisture-related hoof problems. Ensuring clean and dry bedding, especially during wet weather, further supports hoof health.

Maintaining an optimal body condition is integral to preventive care. Adequate nutrition supports overall health and disease resilience, including access to high-quality forage and balanced concentrates. Monitoring body condition scores allows herders to make informed adjustments to feeding programs and identify potential issues, such as overfeeding or underfeeding.

Routine dental care is essential for goats to maintain proper chewing and digestion. Monitoring signs of dental problems, such as overgrowth or uneven wear, helps prevent issues impacting feed intake and nutrient absorption. Providing appropriate dental chews or blocks and a fiber-rich diet supports dental health.

Reproductive care is a crucial aspect of preventive measures, particularly for breeding goats. Maintaining optimal reproductive health involves regular assessments of breeding bucks and does. Ensuring the reproductive soundness of bucks, including evaluating testicular health and libido, contributes to successful breeding outcomes. Monitoring the reproductive cycle of does, assessing overall health, and addressing any potential issues enhance breeding efficiency.

Quarantine protocols for new additions to the herd are essential preventive measures to avoid introducing infectious diseases. Isolating new goats for a period of observation, typically 30 days, allows herders to monitor for signs of illness and prevent the potential spread of contagious diseases to the existing herd.

Biosecurity practices, including limiting exposure to visitors and other animals, help mitigate the risk of disease introduction. Implementing sanitation measures, such as disinfecting equipment and footwear, contributes to a healthy and disease-resistant herd. Awareness of biosecurity principles and collaboration with a veterinarian to develop and implement effective biosecurity protocols enhance overall herd health.

Regular health monitoring and record-keeping are integral components of preventive care. Detailed health records allow herders to track vaccinations, deworming schedules, reproductive histories, and any health issues or treatments. These records provide a valuable resource for making informed decisions, assessing overall herd health trends, and facilitating communication with veterinarians.

Educating oneself about prevalent diseases in the region and staying informed about emerging health concerns is an ongoing aspect of preventive care. Participating in educational programs and workshops and collaborating with veterinary professionals contribute to a herder's ability to proactively manage and address health challenges.

In conclusion, establishing a comprehensive vaccination schedule and implementing preventive care measures are essential components of responsible goat husbandry. Core vaccinations, strategic deworming programs, hoof care, dental care, reproductive health assessments, and biosecurity measures contribute to overall herd health and productivity. Regular health monitoring, record-keeping, and ongoing education further enhance a herder's ability to manage and address health challenges proactively. By prioritizing preventive care, herders can foster a resilient and thriving herd, exemplifying the principles of sustainable and responsible goat farming.

Handling Common Health Issues

Handling common health issues is integral to responsible goat husbandry, as goats can be susceptible to various ailments that may impact their well-being and productivity. Respiratory problems, such as pneumonia, can arise due to environmental factors, including changes in weather, inadequate ventilation, or exposure to drafts. Symptoms may include coughing, nasal discharge, and labored breathing. Prompt veterinary intervention and isolation of affected individuals to prevent disease spread are crucial for managing respiratory infections. Adequate ventilation in barns and shelters, proper bedding, and minimizing exposure to drafts contribute to preventing respiratory issues in goats.

Both internal and external parasitic infestations are common health challenges in goats. Gastrointestinal worms, such as Haemonchus contortus, can lead to conditions like barber pole worm infestations, resulting in anemia, weight loss, and lethargy. Strategic deworming programs, guided by regular fecal testing, help manage internal parasites. External parasites, including lice and mites, can cause skin irritation and discomfort. Topical treatments and maintaining clean living conditions are effective measures for controlling external parasites in goats.

Hoof-related problems, such as foot rot, can affect goat health, particularly in moist environments. Foot rot is characterized by lameness, swelling, and a foul odor. Prompt attention to affected individuals, along with proper hoof trimming, cleaning, and topical treatments, aids in managing and preventing the spread of foot rot within the herd. Providing well-drained living areas, avoiding excessive moisture, and practicing good sanitation further prevent hoof issues.

Metabolic disorders, such as ketosis and pregnancy toxemia, can impact goats, especially during critical periods such as pregnancy and lactation. Ketosis, often seen in dairy goats, results from an energy imbalance and is characterized by symptoms such as lethargy, poor appetite, and acetone-smelling breath. Pregnancy toxemia, common in late-pregnant does, manifests as weakness, confusion, and difficulty standing. Adjusting the diet to meet the increased energy demands during these periods, providing proper nutrition, and monitoring body condition scores help prevent metabolic disorders.

Digestive issues, including bloat and enterotoxemia, can affect goats, particularly those with sudden changes in diet. Bloat, characterized by a distended abdomen, can result from the rapid consumption of lush forage or legumes. Ensuring gradual transitions to new diets and avoiding sudden access to high-risk forages prevent bloat. Enterotoxemia, caused by the bacterium Clostridium perfringens, leads to premature death and is associated with high-carbohydrate diets. Vaccination against enterotoxemia and careful dietary management help mitigate the risk of digestive issues.

Urinary calculi, or blockages in the urinary tract, are common in male goats and can lead to severe discomfort and potential kidney damage. They provide a balanced diet with appropriate calcium and phosphorus levels and access to clean water, which aids in preventing urinary calculi. Monitoring water intake and ensuring a diet conducive to urinary health are essential preventive measures.

Reproductive issues, including dystocia (difficulty kidding) and retained placenta, can occur during the kidding process. Dystocia may result from factors such as improper positioning of the kid or a mismatch in size between the kid and the birth canal. Prompt veterinary assistance and careful monitoring during kidding help manage dystocia.

Retained placenta, where the afterbirth is not expelled within a few hours after kidding, requires veterinary attention to prevent infections and complications.

Nutritional deficiencies, such as essential minerals like selenium and copper, can impact goat health. Selenium deficiency can particularly lead to conditions like white muscle disease, characterized by muscle weakness and difficulty standing. Providing mineral supplements and ensuring that the diet meets the specific nutritional needs of goats helps prevent deficiencies. However, careful attention should be paid to avoid over-supplementation, as excessive intake of certain minerals can lead to toxicity.

Infectious diseases, such as contagious caprine pleuropneumonia (CCPP) and caseous lymphadenitis (CLA), can pose significant threats to goat herds. CCPP, caused by Mycoplasma capricolum subspecies capripneumoniae, affects the respiratory system and can lead to severe pneumonia. CLA, caused by the bacterium Corynebacterium pseudotuberculosis, results in abscess formation in lymph nodes. Implementing biosecurity measures, including quarantine protocols for new additions, helps prevent the introduction of infectious diseases. Vaccination programs targeting prevalent diseases in the region further contribute to disease prevention.

Heat stress is a concern, especially in hot climates, and can lead to hyperthermia and dehydration. Providing shade, adequate ventilation, and access to clean and cool water are essential for preventing heat stress in goats. Avoiding strenuous activities during the hottest parts of the day and implementing cooling measures, such as misting systems, further mitigate heat-related issues.

Handling common health issues requires a proactive approach to herd management. Regular health checks, monitoring for signs of illness, and collaboration with a veterinarian contribute to early detection and effective management of health challenges. Timely interventions, such as administering medications, implementing quarantine measures, and adjusting diets, help address specific health issues and prevent their escalation within the herd. Overall, a combination of preventive measures, attentive care, and a swift response to emerging health concerns fosters the resilience and well-being of a goat herd, exemplifying the principles of responsible and sustainable goat farming.

CHAPTER VI

Breeding Basics

The Ins and Outs of Goat Reproduction

Understanding the intricacies of goat reproduction is crucial for successful and responsible goat husbandry. Goats are prolific breeders with distinct reproductive patterns. Female goats, known as does, typically reach sexual maturity at around six to nine months of age, although this can vary based on breed and individual development. The onset of puberty in males, or bucks, occurs a bit later, usually between seven and ten months. However, it's crucial to note that breeding should be delayed until they are physically and mentally mature to handle the demands of reproduction.

The reproductive cycle of a female goat is known as the estrous cycle, with an average duration of 18 to 21 days. This cycle consists of distinct phases: Proestrus, estrus, metestrus, and diestrus. Behavioral changes, including restlessness and increased vocalization, characterize Proestrus. Estrus, the fertile period, typically lasts about 24 to 48 hours and is marked by receptivity to the buck, a swollen vulva, and a clear or slightly bloody discharge. The subsequent metestrus and diestrus phases involve the preparation of the uterus for pregnancy or returning to a non-receptive state if mating hasn't occurred.

The timing of breeding is critical to ensure successful pregnancies. While goats are considered seasonally polyestrous, meaning they cycle throughout the year, some breeds exhibit seasonal breeding patterns influenced by changes in day length. This is particularly true for breeds in higher latitudes with distinct seasonality. For displaying seasonal breeding, the most fertile period, or the peak of the breeding season, usually occurs during the fall.

Detection of estrus signs is essential for accurate breeding timing. Bucks can be highly effective in detecting estrus in does, and visual or physical contact between the sexes often triggers the onset of estrus. Alternatively, using teaser bucks, vasectomized males can help identify does in estrus without the risk of actual breeding. Observation of mounting behavior, tail flagging, and sniffing are common indicators of estrus in does.

Artificial insemination (AI) is a widespread technique in goat reproduction, offering advantages such as genetic diversity and disease control. However, successful AI requires proper timing, precise technique, and knowledge of an individual's reproductive cycles. AI can be particularly beneficial for breeders aiming to introduce specific genetic traits or overcome geographical constraints in accessing quality breeding stock.

Once bred, the gestation period for goats is approximately 150 days, although it can vary slightly based on factors such as breed, nutrition, and overall health. Adequate nutrition during pregnancy is vital to support the developing fetus and ensure a healthy kidding process. Monitoring pregnant women for signs of distress, providing proper shelter, and offering supplemental nutrition contribute to successful pregnancies.

The kidding or parturition process is a critical phase in goat reproduction. Signs of impending kidding include udder enlargement, relaxation of the pelvic ligaments, and behavioral changes such as restlessness and nesting behavior. Some may isolate themselves, while others may seek the herd's comfort. Providing a clean and quiet kidding environment and monitoring closely during labor is essential. In cases where assistance is needed, such as a malpresentation, breeders should be prepared to intervene or seek veterinary aid promptly.

After kidding, the postpartum period involves the doe's recovery and the care of the newborn kids. Ensuring kids receive colostrum, the nutrient-rich first milk is crucial for their immune system development. Adequate nutrition for lactating supports milk production and the overall health of both the doe and the kids. Implementing proper management practices, including ear tagging or banding for identification, disbudding if desired, and regular health checks, contributes to the well-being of the newborn kids.

Bucks play a significant role in goat reproduction, influencing the genetic traits passed on to future generations. Proper management of breeding bucks involves ensuring their reproductive health, addressing any potential issues, and monitoring their breeding activities. Maintaining optimal body condition, providing a balanced diet, and avoiding overuse to prevent exhaustion are essential for buck management. Rotational use of bucks and strategic breeding planning help maintain genetic diversity within the herd.

Understanding the reproductive dynamics of goats also involves recognizing reproductive disorders and addressing them promptly. Conditions such as infertility, repeat breeding, or difficulty in kidding may necessitate veterinary intervention to identify underlying causes. Regular reproductive health checks, including assessments of reproductive organs, sperm quality in bucks, and monitoring of breeding performance, contribute to proactive management and early detection of potential issues.

Reproductive health management extends to disease prevention, with vaccinations and biosecurity practices playing a vital role. Preventing the spread of infectious diseases, particularly those affecting the reproductive system, ensures the overall health and productivity of the herd. Collaborating with veterinarians to establish vaccination schedules, implement biosecurity measures, and address emerging health concerns supports responsible goat reproduction.

In conclusion, navigating the ins and outs of goat reproduction requires a comprehensive understanding of the estrous cycle, breeding techniques, gestation, parturition, and postpartum care. Responsible breeding practices involve strategic timing, proper management of breeding bucks, and attentive care during kidding. Monitoring reproductive health, addressing potential issues promptly, and implementing disease prevention measures contribute to successful and sustainable goat reproduction. By embracing a holistic approach to goat reproduction, breeders can foster a resilient and thriving herd that exemplifies the principles of responsible and ethical goat husbandry.

Planning and Managing Breeding Seasons

Planning and managing breeding seasons is a critical aspect of responsible goat husbandry, influencing the herd's overall health, productivity, and sustainability. The choice of breeding season is a decision that requires careful consideration of various factors, such as the natural reproductive cycle of goats, environmental conditions, and the specific goals of the breeding program. While goats are considered seasonally polyestrous, meaning they can cycle throughout the year, some breeds exhibit seasonal breeding patterns influenced by changes in day length. Understanding these patterns is critical to strategically planning breeding seasons.

The most common breeding seasons for goats are fall and spring, aligning with the natural reproductive cycle of many breeds. Fall breeding, often called "out-of-season" breeding, capitalizes on the increased fertility of does during the fall months. This approach allows for kidding during the milder spring weather when forage availability is typically abundant. Alternatively, spring breeding aligns with the natural breeding season for many goats, resulting in spring or early summer kiddings. The choice between fall and spring breeding depends on factors such as climate, forage availability, market demand for goat products, and the preferences or goals of the goat breeder.

One of the primary considerations in planning breeding seasons is the environmental factor. Does experience increased fertility in response to decreasing day length, commonly observed during the fall months. Breeders aiming for fall kiddings strategically plan to introduce the bucks to the does approximately three to four months before the desired kidding period. This timing aligns with the natural cycle of goats, optimizing reproductive success. Additionally, fall kiddings offer the advantage of avoiding harsh winter conditions during the vulnerable early weeks of the kids' lives.

Spring breeding, conversely, corresponds to the natural breeding season for many goat breeds. As the days lengthen and temperatures rise, they naturally enter estrus, signaling their receptivity to breeding. Planning for spring kiddings involves introducing the bucks to the does during winter, allowing for kids' birth in the more favorable spring and early summer weather. Spring kiddings provide the advantage of ample forage availability during the lactation period, supporting optimal nutrition for dogs and their offspring.

The goals of the breeding program significantly influence the choice of breeding season. For breeders focused on meat production, aligning kiddings with periods of abundant forage, such as spring, may optimize weight gain and the kids' overall health. For dairy producers, strategic planning to ensure fresh forage availability during peak lactation is crucial. They understand the market demand for goat products, whether meat, milk, or fiber, which helps breeders tailor their breeding seasons to meet consumer preferences.

Managing breeding seasons also involves considerations related to reproductive performance and efficiency. Doe management influences reproductive success, including body condition, health, and age. Those in good body condition are more likely to conceive and maintain pregnancies successfully. Proper nutrition during the breeding season, which may include adjustments to the does' diet to meet increased energy demands, supports reproductive health. Additionally, monitoring the do's age is essential, as younger does may require more attention to ensure successful pregnancies.

Buck management is equally critical in planning and managing breeding seasons. Bucks should be in good physical condition, free from health issues, and have adequate libido for successful mating. Proper nutrition, parasite control, and regular health checks contribute to the reproductive fitness of bucks. Rotational use of bucks is a common practice, preventing exhaustion and maintaining genetic diversity within the herd. Breeders may also conduct breeding soundness evaluations on bucks to ensure their reproductive fitness.

Implementing a controlled breeding program involves synchronizing the estrous cycles of does, allowing for a more concentrated kidding period. This approach benefits commercial operations aiming to streamline kidding management and maximize labor efficiency. Controlled breeding often utilizes hormonal interventions, such as prostaglandin treatments, to synchronize the breeding readiness of does. However, it requires precise timing and careful observation to optimize success rates.

Monitoring reproductive health is an ongoing aspect of managing breeding seasons. Regular health checks, including assessments of reproductive organs, detecting estrus signs, and monitoring breeding activities, contribute to proactive management and early detection of potential issues. Identifying and promptly addressing reproductive disorders is crucial for maintaining the overall health and productivity of the herd.

Biosecurity measures play a role in managing breeding seasons, especially when introducing new breeding stock. Quarantine protocols for new additions, including isolation and health checks, help prevent the potential spread of infectious diseases within the herd. Vaccination programs tailored to the specific needs of the breeding season contribute to disease prevention and overall herd health.

In conclusion, planning and managing breeding seasons is a multifaceted endeavor that requires a thoughtful approach. Considerations such as environmental factors, breeding goals, reproductive performance, and biosecurity measures all play integral roles in determining the timing and success of breeding seasons. Whether aiming for fall or spring kiddings, aligning breeding seasons with the natural reproductive cycles of goats is fundamental. By adopting strategic breeding practices, attending to the health and nutritional needs of does and bucks, and implementing biosecurity measures, breeders can foster a resilient and productive herd that aligns with responsible and sustainable goat farming principles.

Caring for Pregnant and Nursing Does

Caring for pregnant and nursing is a critical aspect of responsible goat farming, ensuring the health and well-being of the does and their offspring. Pregnancy imposes increased nutritional demands on does, and attentive care during this period contributes to successful pregnancies and optimal kidding outcomes. Adequate nutrition is paramount, with a focus on providing a balanced diet that meets the specific needs of pregnant women. This includes a combination of high-quality forage, concentrates, and mineral supplements to address the increased energy and nutrient requirements during gestation.

Monitoring the body condition of pregnant women is essential, as maintaining an optimal body condition score contributes to overall reproductive health. Those who enter the kidding period with a proper body condition are likelier to deliver healthy kids and exhibit good mothering instincts. Adjustments to the feeding program may be necessary as the pregnancy progresses, ensuring that it receives the nutrients needed to support fetal development and lactation.

Access to clean and fresh water is a fundamental aspect of caring for pregnant women. Adequate hydration is crucial for the health of both the does and their developing fetuses. Monitoring water intake, especially during hot weather, helps prevent dehydration and supports optimal reproductive performance.

Providing pregnant women with a suitable and comfortable environment is integral to their well-being. Adequate shelter from harsh weather conditions, proper bedding to minimize stress, and sufficient space to move around contribute to a stress-free pregnancy. Reducing stress is particularly important during the final weeks of pregnancy, as stress can trigger premature kidding and negatively impact the health of the doe and the kids.

As kidding approaches, vigilant monitoring of pregnant women becomes imperative. Signs of impending labor include restlessness, nesting behavior, udder enlargement, and relaxation of the pelvic ligaments. Familiarizing oneself with these signs allows breeders to intervene promptly or seek veterinary assistance in case of complications. Providing a clean and quiet kidding environment and access to a private space further supports a smooth kidding process.

Postpartum care is a continuation of the attentive care provided during pregnancy. Ensuring newborn kids receive colostrum, the nutrient-rich first milk is critical for their immune system development. Observing the mothering behavior of does helps identify any potential issues, such as inadequate care or rejection of kids. In cases where parents are reluctant to nurse, supplementary feeding may be necessary to ensure the kids receive adequate nutrition.

Managing the nutritional needs of nursing involves providing a diet that supports lactation. The demands of milk production require a balanced diet with increased energy and protein content. Offering high-quality forage, supplemented with concentrates and mineral supplements, helps meet the nutritional requirements of lactating. Monitoring body condition scores during lactation allows adjustments to the feeding program to prevent excessive weight loss and maintain overall health.

Regular health checks for pregnant and nursing are essential for early detection and intervention of health issues. Monitoring for signs of distress, assessing body condition, and observing behavioral changes contribute to proactive management. Addressing any health concerns promptly, whether related to reproductive issues, parasite infestations, or other ailments, helps ensure the well-being of the does and their offspring.

Biosecurity measures play a role in caring for pregnant and nursing nurses, especially when introducing new breeding stock. Quarantine protocols for new additions, including health checks and isolation, help prevent the potential spread of infectious diseases within the herd. Vaccination programs tailored to the specific needs of pregnant and lactating contribute to disease prevention and overall herd health.

Beyond the physical aspects of care, providing pregnant women and nurses with a stress-free and calm environment is crucial. Goats are sensitive to changes in their surroundings, and minimizing disturbances contributes to their overall well-being. Gentle handling, routine care, and a consistent daily routine help reduce stress and promote a positive environment during these critical periods.

In conclusion, caring for pregnant women and nursing are multifaceted responsibilities encompassing nutritional management, attentive monitoring, and proactive health care. Adequate nutrition, access to clean water, proper shelter, and a stress-free environment contribute to the well-being of women during pregnancy and lactation. Monitoring for signs of distress, providing postpartum care to newborn kids, and implementing biosecurity measures further enhance responsible goat husbandry. By prioritizing the needs of pregnant and nursing dogs, breeders can foster a healthy and thriving herd, exemplifying the principles of sustainable and ethical goat farming.

CHAPTER VII

Kid Care: From Birth to Adulthood

Welcoming New Kids to the Herd

Welcoming new kids to the herd is a joyous and pivotal moment in the life of a goat farm, marking the beginning of a new generation and the continuation of the herd. The arrival of kids, born through careful breeding and attentive management, is a testament to successful reproduction and the herd's overall health. As newborns are particularly vulnerable, creating a supportive and nurturing environment is crucial for their well-being and development.

The intake of colostrum defines the first moments of a kid's life, the nutrient-rich first milk produced by the doe. Colostrum is essential for the kid's immune system development, providing antibodies and nutrients that protect against infections. Ensuring that each newborn receives adequate colostrum within the first few hours of life is fundamental to welcoming new kids. Breeders should closely monitor the nursing behavior, and in cases where a kid is reluctant to nurse, or the mother exhibits rejection, intervention may be necessary. Supplementary feeding with colostrum substitutes or goat milk can guarantee the kid receives the vital nutrients needed for a healthy start.

Maintaining a warm and clean environment is crucial for the well-being of newborn kids. Newborns are susceptible to chilling, and ensuring access to a draft-free, warm shelter is essential. Bedding materials, such as straw or hay, should be clean and dry to prevent health issues and promote a comfortable resting space for the kids. Adequate ventilation is necessary to avoid the buildup of humidity, which can contribute to respiratory problems. Heat lamps or kid jackets can be used for added warmth, especially during colder seasons.

Early identification and tagging of newborn kids are essential for herd management and record-keeping. Ear tagging or banding is commonly employed for easy identification, and this practice aids in tracking lineage, health records, and other essential information. This early tagging ensures that each kid is accounted for and facilitates effective monitoring and care throughout their lives.

Managing the transition from the kidding area to the larger herd environment requires careful consideration. While the initial weeks are critical for bonding between the mother and kid, socialization with other herd members is equally important. Introducing kids to the larger herd gradually, allowing them to interact with other goats while maintaining proximity to their mothers, promotes a smooth transition. Monitoring the dynamics within the herd and intervening in cases of aggression or bullying ensures the safety and well-being of the newborns.

Proper nutrition for growing kids is paramount to their development and overall health. Offering a balanced diet is crucial as they transition from colostrum to milk and eventually solid food. Milk replacers can be introduced when weaning is initiated, typically around two to three months of age. The composition of milk replacers should match the nutritional needs of growing kids, ensuring they receive the necessary vitamins and minerals for optimal development.

Access to high-quality forage and appropriate concentrates supports the nutritional requirements of growing kids. Monitoring the body condition of kids and adjusting their diet accordingly is essential to prevent issues such as overfeeding or underfeeding. A well-balanced diet contributes to their physical development and lays the foundation for future productivity and reproductive health.

A comprehensive newborn healthcare program is vital to prevent and address common issues. Vaccinations, deworming, and hoof trimming are integral components of kid care. Establishing a vaccination schedule guided by veterinary advice protects against prevalent diseases and contributes to the herd's overall health. Deworming programs, tailored to the specific needs of kids, help manage internal parasites that can affect growth and overall well-being. Regular hoof trimming, initiated as needed, prevents issues such as overgrowth and misshapen hooves.

Monitoring for signs of illness and addressing health concerns promptly is essential for kid care. Common health issues in newborn kids include respiratory infections, digestive problems, and parasitic infestations. Respiratory issues may arise from environmental factors or infectious agents, and prompt veterinary attention is necessary for effective management. Monitoring for signs of scours or diarrhea aids in the early detection of digestive issues, and adjustments to the diet may be required. Regular fecal testing helps identify and manage internal parasites, ensuring the kids' overall health.

Kid care extends beyond physical health to include behavioral and social aspects. Observing the social dynamics within the herd and addressing any signs of bullying or exclusion is crucial for the well-being of kids. Providing ample opportunities for play and exploration supports their cognitive and physical development. Creating a positive and stimulating environment encourages the natural behaviors of kids, fostering a sense of security and contentment within the herd.

As kids grow and mature, preparing them for their roles within the herd becomes a gradual process. Introducing them to handling and routine care practices, such as hoof trimming and vaccinations, helps acclimate them to human interaction. This early exposure facilitates easier management in the future and contributes to the overall docility of the herd.

Welcoming new kids to the herd is a celebration of life and a commitment to responsible and sustainable goat farming. By providing a nurturing environment, addressing their nutritional needs, implementing health care programs, and fostering positive social dynamics, breeders contribute to the well-being and future productivity of the entire herd. The successful integration of newborn kids into the larger herd is a testament to effective management practices and exemplifies the principles of ethical and attentive goat husbandry.

Essential Care During the First Weeks

Ensuring essential care during the first weeks of a kid's life is a cornerstone of responsible goat husbandry, shaping the foundation for their growth, health, and overall well-being. The initial weeks are a critical period, demanding close attention to various aspects of care, from nutrition and hydration to environmental considerations and health monitoring.

First and foremost, the intake of colostrum, the nutrient-rich milk the doe produces, is a pivotal aspect of early care. Colostrum is a source of vital nutrients and contains antibodies crucial for bolstering the kid's immune system. Ensuring that each newborn receives colostrum within the first few hours of life is non-negotiable. This initial feeding provides essential antibodies that protect against infections, offering a crucial immunity boost during the vulnerable early days of a kid's life.

Maintaining a warm and clean environment is paramount during the first weeks. Newborn kids are particularly susceptible to chilling, and exposure to cold temperatures can lead to health complications. Providing a draft-free and warm shelter, especially during colder seasons, is essential for their well-being. Proper bedding, composed of clean and dry materials such as straw or hay, offers comfort and helps prevent health issues. Additionally, adequate ventilation is necessary to ensure fresh air circulation and reduce the risk of respiratory problems.

Hydration is another fundamental consideration during the first weeks of a kid's life. While colostrum provides essential fluids, monitoring the kid's water intake as they consume water independently is crucial. Clean and fresh water should be readily available to support proper hydration, contributing to overall health and preventing dehydration.

Early identification and tagging are significant in managing and tracking the herd. Ear tagging or banding is commonly employed for identification, allowing breeders to monitor lineage, health records, and other essential information throughout the kid's life. This practice aids in effective record-keeping and facilitates herd management, ensuring that each kid is accounted for and contributing to the overall success of the breeding program.

Observing the mothering behavior of does is essential during the first weeks, as it provides insights into the well-being of the do and their kids. While most exhibit natural mothering instincts, some may require additional attention or support. In cases where a doe shows rejection or reluctance to nurse, breeders should be prepared to intervene. Supplementary feeding with colostrum substitutes or goat milk may be necessary to ensure the kid receives adequate nutrition for healthy growth.

The transition from the kidding area to the larger herd environment is a carefully managed process during the first weeks. While bonding between the mother and kid is essential, socialization with other herd members is equally important. Introducing kids to the larger herd gradually allows them to interact with other goats while maintaining proximity to their mothers. This gradual approach promotes a smooth transition and minimizes stress, ensuring the kids feel secure in their new environment.

Proper nutrition is a critical component of essential care during the first weeks. As kids transition from colostrum to milk and eventually solid food, ensuring a balanced diet is crucial for their development. Adequate nutrition supports growth, immune system function, and overall health. Milk replacers can be introduced when weaning is initiated, typically around two to three months of age. Monitoring body condition scores during this period allows adjustments to the feeding program to prevent overfeeding or underfeeding.

Implementing a comprehensive health care program during the first weeks is vital to prevent and address common issues. Vaccinations, deworming, and hoof trimming are integral to early kid care. Establishing a vaccination schedule guided by veterinary advice protects against prevalent diseases and contributes to overall herd health. Deworming programs, tailored to the specific needs of kids, help manage internal parasites that can affect growth and overall well-being. Regular hoof trimming, initiated as needed, prevents issues such as overgrowth and misshapen hooves.

Monitoring for signs of illness and addressing health concerns promptly is essential during the first weeks. Common health issues in newborn kids include respiratory infections, digestive problems, and parasitic infestations. Respiratory issues may arise from environmental factors or infectious agents, and prompt veterinary attention is necessary for effective management. Monitoring for signs of scours or diarrhea aids in the early detection of digestive issues, and adjustments to the diet may be required. Regular fecal testing helps identify and manage internal parasites, ensuring the kids' overall health.

Beyond physical health, attending to behavioral and social aspects during the first weeks is crucial. Observing the social dynamics within the herd and addressing any signs of bullying or exclusion ensures the well-being of kids. Providing ample opportunities for play and exploration supports their cognitive and physical development.

Creating a positive and stimulating environment encourages the natural behaviors of kids, fostering a sense of security and contentment within the herd.

In conclusion, essential care during the first weeks of a kid's life is a comprehensive and multifaceted responsibility. From ensuring the intake of colostrum to managing the transition to the larger herd, providing proper nutrition, implementing health care programs, and fostering positive social dynamics, breeders contribute to their herd's overall well-being and future productivity. The success of early care practices is reflected in the health, vitality, and resilience of the kids as they grow into productive and thriving members of the herd, exemplifying the principles of ethical and attentive goat husbandry.

Growing Healthy Kids: Challenges and Solutions

Growing healthy kids poses opportunities and challenges in goat farming, as this phase is crucial for their overall development and eventual contribution to the herd. Challenges during growth can manifest in various forms, from nutritional imbalances and parasitic infestations to behavioral issues and infectious diseases. Addressing these challenges requires a comprehensive approach encompassing nutrition, health care, and attentive management practices.

Nutrition plays a central role in kids' healthy growth, and challenges in this area often revolve around providing a balanced diet that meets their evolving needs. While milk is the primary source of nutrition during the early weeks, weaning introduces the challenge of transitioning to solid food. Ensuring access to high-quality forage, supplemented with concentrates and mineral supplements, is essential to support the nutritional requirements of growing kids. Monitoring body condition scores and adjusting the diet accordingly is crucial to prevent issues such as overfeeding or underfeeding, which can have implications for the long-term health and productivity of the growing goats.

Parasitic infestations represent another common challenge during the growth phase of kids. Internal parasites, such as worms, can impact goats' overall health and growth. Implementing a deworming program tailored to the specific needs of growing kids helps manage these parasites effectively. Regular fecal testing, guided by veterinary advice, aids in identifying and addressing parasitic infestations promptly. However, overreliance on dewormers without considering factors such as pasture management and rotational grazing can contribute to the development of parasite resistance, requiring a balanced and strategic approach.

Behavioral challenges may arise as kids grow and interact within the herd. Aggression or bullying from older goats, especially during the integration process, can impact the well-being of kids being raised. Observing the social dynamics within the herd and intervening in cases of aggression helps ensure a positive and supportive environment for all members. Providing ample space, appropriate shelters, and opportunities for play and exploration contributes to a harmonious herd dynamic, fostering the social and cognitive development of the growing goats.

Infectious diseases pose a potential threat to the health of growing kids, necessitating proactive health care measures. Vaccination programs, tailored to the herd's specific needs and guided by veterinary advice, play a crucial role in preventing common diseases. Establishing a vaccination schedule and ensuring that all kids receive vaccinations supports overall herd health. Additionally, biosecurity measures, such as quarantining new additions and practicing proper sanitation, help prevent the spreading of infectious diseases within the herd.

Maintaining hoof health is an overlooked but critical aspect of growing kids' overall well-being. Hoof trimming, initiated as needed, prevents overgrowth and misshapen hooves. Regular checks for signs of lameness or hoof abnormalities allow breeders to promptly address potential issues, ensuring the growing goats' mobility and comfort.

The growth phase is also characterized by the development of reproductive maturity, especially in does. While this signifies a natural progression, premature breeding can pose challenges, particularly in younger children. Managing breeding programs and ensuring that dogs are physically and mentally mature before breeding helps prevent potential complications and supports the long-term reproductive health of the herd.

External factors, such as climatic conditions and environmental stressors, can also impact the health and growth of kids. Harsh weather, inadequate shelter, or extreme temperatures can contribute to stress and compromise the immune system. Providing well-designed shelters and implementing management practices that mitigate the impact of environmental stressors contribute to growing goats' overall resilience.

Genetic considerations play a significant role in the growth and development of kids. Breeders aiming for specific traits or characteristics should carefully select breeding pairs to achieve desired outcomes. Genetic diversity within the herd helps maintain resilience and adaptability to various environmental conditions.

Solutions to these challenges involve a holistic and proactive approach to goat farming. Implementing rotational grazing practices, managing pasture health, and practicing strategic parasite control contribute to a balanced and sustainable nutritional environment for growing kids. Regular health checks, vaccinations, and deworming programs support their overall health and prevent the occurrence of common diseases.

Providing a stimulating and stress-free environment, alongside proper behavioral management, fosters positive social interactions and contributes to the cognitive development of growing goats.

Balancing the nutritional needs of growing kids involves formulating diets that meet their evolving requirements. Collaborating with a veterinarian to establish appropriate vaccination schedules, conducting routine health checks, and promptly addressing any health concerns contribute to the overall well-being of the herd. Genetic selection and responsible breeding practices ensure the transmission of desirable traits and maintain the vigor of the herd.

In conclusion, growing healthy kids involves navigating various challenges with a comprehensive and proactive mindset. From nutritional considerations and parasite control to behavioral management and disease prevention, each aspect contributes to the overall health and productivity of the growing goats. By addressing these challenges with attentive care, sound management practices, and a commitment to genetic diversity, breeders can foster a generation of healthy, resilient, and thriving goats that exemplify the principles of responsible and sustainable goat farming.

CHAPTER VIII

Training and Handling Techniques

Establishing Trust with Your Herd

Establishing trust with your herd is a cornerstone of successful goat husbandry, fostering a harmonious and cooperative relationship between humans and these intelligent, social animals. Building trust is a gradual process that requires patience, consistency, and a deep understanding of goat behavior. By investing time and effort into developing a trust-based bond, herders can create an environment where goats feel secure, leading to easier management, reduced stress, and enhanced overall well-being.

One of the primary foundations of trust-building is spending quality time with the herd. Goats are naturally curious, and social animals and consistent human presence allows them to become familiar with their caretakers. Spending time in the pasture, observing their behaviors, and engaging in non-threatening activities contribute to building positive associations with human presence. Sitting quietly and allowing goats to approach at their own pace helps establish a sense of safety and reduces fear.

Consistent and gentle handling is paramount in the trust-building process. Goats are sensitive to changes in their environment and can be easily startled. Approach them calmly and avoid sudden movements or loud noises that may cause anxiety. Regular, gentle handling, such as petting and scratching, reinforces positive associations with human touch. This tactile interaction, especially when initiated in a non-threatening manner, helps goats become accustomed to human contact and gradually builds trust.

Feeding time presents an excellent opportunity to reinforce positive associations with humans. Offering treats or supplemental feed while spending time with the herd encourages them to associate humans with positive experiences. This positive reinforcement can be particularly effective during initial interactions or when introducing new goats to the herd. However, it's crucial to ensure that treats are healthy and appropriate for goats, avoiding excessive or unbalanced feeding.

Creating a consistent routine contributes to the establishment of trust. Goats are creatures of habit, and they feel more secure when they know what to expect. Consistent feeding times, routine health checks, and predictable interactions help goats anticipate and understand human activities. This predictability reduces stress and fosters a sense of security, contributing to the development of trust.

Understanding and respecting the natural behaviors of goats is essential in building trust. Goats have a hierarchical social structure, and respecting their personalities and preferences goes a long way in earning their trust. Observing their body language, recognizing signs of stress or discomfort, and adjusting your interactions demonstrate an understanding that fosters trust. Additionally, avoiding aggressive behavior and maintaining a calm and assertive demeanor contribute to positive interactions with the herd.

Patience is a virtue in the process of trust-building with goats. Some individuals may be more reserved or cautious, requiring more time to acclimate to human interaction. Forcing interactions or pushing boundaries can erode trust and result in fearful behavior. It's crucial to allow each goat to progress at its own pace, offering constant reassurance and positive reinforcement.

Training goats using positive reinforcement techniques can be an effective way to build trust and enhance communication. Associating specific cues with positive experiences, such as treats or praise, allows goats to understand and respond to commands. This can be useful in situations like leading goats on a leash, facilitating health checks, or managing the herd. Positive reinforcement training builds trust and creates a cooperative and responsive herd.

Creating a secure and goat-friendly environment is integral to trust-building. Adequate shelter, clean water, and appropriate forage contribute to the overall well-being of the herd. Providing space for goats to exhibit natural behaviors, such as climbing, browsing, and resting, enhances their contentment and trust. Well-designed and maintained facilities, including handling areas and shelters, contribute to stress-free interactions and positive associations with human presence.

Consistent and clear communication is essential in building trust with goats. Goats are perceptive animals that respond to vocal cues, body language, and tone of voice. Using a calm and reassuring tone and consistent cues helps goats understand expectations and fosters a sense of security. Avoiding sudden movements or loud noises during interactions contributes to a positive communication environment.

Building trust with goats is an ongoing process that extends beyond physical interactions. Respect for their autonomy and allowing them to express their natural behaviors contribute to a positive relationship. Creating a herd dynamic that minimizes stress, avoids aggressive behavior, and supports positive social interactions enhances the overall trust within the group. Understanding the unique personalities of individual goats and adapting your approach accordingly contributes to a more nuanced and trusting relationship.

In conclusion, establishing trust with your herd is fundamental to responsible and ethical goat husbandry. Through consistent and positive interactions, understanding their natural behaviors, and creating a secure and goat-friendly environment, herders can build a foundation of trust that enhances the well-being of the goats and facilitates effective management. Trust- building is an investment that pays dividends in terms of easier handling, reduced stress, and a positive and cooperative herd dynamic. Ultimately, a relationship built on trust exemplifies humane and attentive goat care principles, creating an environment where humans and goats thrive together.

Basic Training Commands

Basic training commands are invaluable for establishing clear communication and fostering a cooperative relationship between humans and goats. Training goats enhances their manageability and contributes to their overall well-being and the efficiency of herd management. Implementing basic training commands involves patience, consistency, and an understanding of goat behavior. Herders can create a more harmonious and responsive herd by incorporating these commands into the daily routine.

One fundamental command is "Come" or "Here,"

encouraging goats to approach the handler when called. Teaching goats to respond to their names or specific cues helps establish a connection and reinforces positive interactions. This command is handy during feeding times, health checks, or moving the herd to different areas. Consistent use of the "Come" command and positive reinforcement such as treats or praise reinforces the association between the command and a positive outcome, encouraging goats to respond willingly.

"Stay" or "Wait" is another essential command that aids in controlling the movement of goats. Teaching goats to stand still or wait on command is useful during health checks, grooming, or navigating through gates. This command is particularly beneficial when maintaining a specific position is necessary. Consistent repetition, gentle guidance, and positive reinforcement help goats understand and comply with the "Stay" command, contributing to a more controlled and cooperative herd.

"Back" or "Move Back" is a command that facilitates backward movement. Teaching goats to step back on cue is valuable when navigating through tight spaces, during hoof trimming, or when avoiding over-crowding at feeding stations. Consistent use of the "Back" command and gentle pressure and release help goats understand the desired movement and reinforce a sense of cooperation. This command is handy for establishing boundaries and maintaining a safe distance between goats and handlers.

"Go" or "Move" is a command that encourages forward movement. Teaching goats to respond to the "Go" command is beneficial during herding, leading goats to different areas, or transitioning between spaces. Consistent command use, accompanied by gentle guidance and positive reinforcement, encourages goats to move forward willingly. This command is handy when directing the herd's movement and is necessary for efficient management.

"Stand" is a command that encourages goats to stand still on all four feet. This command is helpful during health checks, grooming, or when preparing goats for milking. Consistent repetition, gentle guidance, and positive reinforcement contribute to goats understanding and compliance with the "Stand" command. This command is precious when working with individual goats and requires them to remain stable for a specific duration.

"Hoof" or "Lift" is a command that teaches goats to present their hooves for inspection or trimming. Regular hoof care is essential for goat health, and training goats to lift their hooves on command simplifies this process. Consistent use of the "Hoof" command, coupled with positive reinforcement and gentle handling, allows goats to become accustomed to having their hooves examined and trimmed, contributing to overall herd health.

"Quiet" or "Hush" is a command that encourages goats to stop vocalizing. Goats can be pretty vocal, especially during feeding times or when agitated. Teaching goats to respond to the "Quiet" command is beneficial when minimizing noise is necessary. Consistent use of the command, coupled with positive reinforcement, helps goats associate the "Quiet" command with a desirable outcome, contributing to a more peaceful and controlled herd environment.

"Go to Your Pen" or "Go to Shelter" is a command that directs goats to a specific location. Teaching goats to respond to this command is helpful for herd management, especially during adverse weather conditions or when goats need to be separated for various reasons. Consistent command use, accompanied by positive reinforcement, encourages goats to move to the designated area willingly, facilitating efficient herd management.

Training goats involves a combination of vocal cues, body language, and positive reinforcement. It's essential to start with one command at a time and gradually introduce additional commands as goats become familiar with the training process. Consistency is vital, and regular, short training sessions are more effective than infrequent, lengthy ones.

Positive reinforcement, such as treats, praise, or gentle petting, is a powerful motivator during training. Goats respond well to rewards, associating the desired behavior with a positive outcome reinforces their willingness to comply with commands. However, it's crucial to use treats judiciously and ensure they are healthy and appropriate for goats.

Understanding the individual personalities and temperaments of goats is essential in tailoring training approaches. Some goats respond more to verbal cues, while others rely more on visual signals or body language. Being attuned to individual goats' preferences and learning styles allows for a more effective and personalized training experience.

Consistent and patient handling during training builds trust and reinforces positive associations with human interaction. Avoiding aggressive behavior, loud noises, or sudden movements during training sessions minimizes stress and contributes to a positive training experience. Goats are intelligent animals, and building a relationship based on trust enhances the success of training endeavors.

In conclusion, incorporating basic training commands into the routine of goat farming contributes to a more cooperative, manageable, and content herd. Teaching commands such as "Come," "Stay," "Back," "Go," "Stand," "Hoof," "Quiet," and "Go to Your Pen" simplifies various aspects of herd management. Through consistent training, positive reinforcement, and a deep understanding of goat behavior, herders can establish a clear line of communication with their goats, fostering a relationship built on trust and cooperation. Basic training commands enhance the efficiency of daily tasks and contribute to the overall well-being of goats, creating a positive and harmonious environment within the herd.

Safe and Effective Handling Practices

Safe and effective handling practices are fundamental to successful goat farming, ensuring the well-being of both the animals and the handlers. Goats, known for their intelligence and sometimes unpredictable behavior, require careful and considerate handling to minimize stress, prevent injuries, and facilitate various management tasks. Whether for routine health checks, grooming, or moving the herd, implementing safe and effective handling practices is essential for creating a positive and cooperative relationship between humans and goats.

One key aspect of safe handling is understanding goat behavior. Goats are prey animals with a strong flight instinct and may react nervously to unfamiliar situations or sudden movements. Handlers should approach goats calmly and avoid making loud noises or sudden gestures that could startle them. Familiarizing oneself with the herd's social structure and individual personalities aids in predicting and managing their reactions, contributing to a smoother handling experience.

Using a calm and confident demeanor during handling instills a sense of security in goats. Handlers should move slowly and deliberately, avoiding any signs of aggression or uncertainty. Goats respond well to assertive but gentle leadership, and a handler's calm presence helps reassure the animals, reducing stress and making them more cooperative during various tasks.

Proper facilities and equipment are crucial for safe handling. Well-designed handling facilities, such as chutes and pens, aid in restraining goats for health checks, vaccinations, or other procedures. Sturdy and well-maintained equipment, such as halters and leads, ensures control without causing harm. Handlers should regularly inspect facilities and equipment for signs of wear or damage to address potential safety concerns promptly.

Approaching goats from their line of vision is a recommended practice for safe handling. Goats have a wide field of view, and coming from the front allows them to see the handler, reducing the likelihood of surprise reactions. Handlers should avoid sudden movements from behind or reaching over the goat's head, as these actions may cause anxiety or resistance.

Understanding the flight zone and point of balance is essential for effective handling. The flight zone is the area around the goat that, when entered by the handler, causes the goat to move away. Handlers can use the flight zone to guide goats in a desired direction without causing stress. The point of balance, located at the goat's shoulder, is crucial for controlling movement. Applying pressure behind the point of balance encourages forward movement, while pressure in front prompts the goat to move backward.

Halter training is a valuable practice for safe handling, especially when working with individual goats. Introducing goats to halters gradually, using positive reinforcement and rewards, helps them become accustomed to the sensation. Once trained, goats can be easily led and controlled using a halter and lead rope, facilitating handling tasks with minimal stress.

When moving goats, it's essential to use low-stress handling techniques. Handlers should avoid excessive prodding or chasing, which can lead to heightened stress levels and make future handling more challenging. Gentle encouragement and an understanding of the herd dynamics allow for a smoother and more controlled movement of goats within the designated areas.

Tailoring handling practices to the specific needs of each goat contributes to a positive experience. Recognizing that individual goats may have different temperaments and responses helps handlers adapt their approach accordingly. Some goats are more receptive to handling, while others require extra patience and reassurance.

Taking the time to build trust with each goat enhances the overall effectiveness of handling practices.

During health checks or veterinary procedures, proper restraint techniques are crucial for the safety of both goats and handlers. Using secure chutes or pens with headlocks helps immobilize goats safely for procedures such as vaccinations, deworming, or hoof trimming. Handlers should be trained in proper restraint techniques to minimize stress and prevent injuries during these necessary tasks.

Preventing overcrowding during handling is essential for maintaining a calm and controlled environment. Goats can become agitated in tight spaces, leading to increased stress and potential injuries. Handlers should ensure that facilities are appropriately sized for the number of goats being handled, allowing for easy movement and minimizing the risk of accidents.

Observing and understanding their social dynamics is crucial when working with goats in a group. Goats have a hierarchical structure, and handlers should be aware of dominant and submissive individuals. This knowledge helps prevent conflicts during handling and ensures that each goat receives proper attention and care without unnecessary stress.

Effective communication between handlers is essential for safe and coordinated handling practices. When multiple handlers are involved, clear signals and consistent commands help prevent confusion and control the herd. Planning and coordination before handling tasks, including assigning specific roles to each handler, contribute to a more efficient and safer handling process.

Adapting handling practices to the age and physical condition of the goats is essential. Kids, pregnant dogs, and older goats may have different needs and responses to handling. Handlers should consider the specific requirements of each group and tailor their approach accordingly. For example, extra care should be taken when handling pregnant women to avoid unnecessary stress that could impact their well-being.

Regular training and familiarization sessions contribute to ongoing positive handling experiences for goats. Handlers can use these sessions to reinforce basic commands, such as "Come" or "Stay," and to maintain the goats' responsiveness to handling. Routine sessions also allow handlers to monitor the health and condition of the goats, making it easier to identify and address any potential issues promptly.

In conclusion, safe and effective handling practices are crucial for goats' well-being and herd management's success. Handlers can create a positive and cooperative relationship with their goats by understanding goat behavior, using calm and confident handling techniques, and employing appropriate facilities and equipment. Tailoring handling practices to the individual needs of each goat, preventing overcrowding, and ensuring clear communication between handlers contribute to a safe, stress-free, and efficient handling environment. As a result, goats become more cooperative during various tasks, fostering a positive relationship between humans and these intelligent and unique animals.

CHAPTER IX

Profitable Ventures: Milk, Meat, and Fiber

Exploring Dairy Goat Farming

Exploring dairy goat farming opens up opportunities for individuals seeking a rewarding venture in agriculture. Dairy goats, known for their docile nature and milk-producing capabilities, have become increasingly popular among small-scale farmers and homesteaders. Dairy goat farming involves a multifaceted approach encompassing breed selection, housing, nutrition, milking practices, and overall herd management. Whether entering the dairy goat industry for personal use or commercial purposes, a comprehensive understanding of the critical aspects involved is crucial for success.

Breed selection is a critical first step in establishing a dairy goat herd. Various goat breeds are recognized for their milk-producing abilities, each with distinct characteristics. Popular dairy goat breeds include the Saanen, Alpine, Nubian, LaMancha, and Nigerian Dwarf. The choice of breed depends on factors such as milk yield, butterfat content, adaptability to the local climate, and specific goals of the farmer. They assess the individual needs and preferences, consider the resources available, and aid in selecting a breed that aligns with the intended scale and objectives of the dairy goat enterprise.

Providing suitable housing and shelter for dairy goats is essential to ensure their well-being and productivity. Goat barns or shelters should be well-ventilated, offering protection from harsh weather conditions and providing proper sanitation. Adequate space per goat, appropriate bedding, and good waste management contribute to a clean and comfortable environment. Additionally, well-designed milking parlors or stanchions are essential for efficient milking practices, providing a controlled setting that minimizes stress for the goats and the farmer.

Nutrition plays a pivotal role in the success of a dairy goat farming operation. Goats require a well-balanced diet that meets their nutritional needs for milk production, growth, and overall health. High-quality forage, such as alfalfa or clover, supplemented with grains, minerals, and vitamins, is a primary diet component. Access to clean water is crucial for hydration and optimal milk production. Balancing the nutritional requirements according to the goats' lactation stage, age, and production goals contributes to their overall health and milk quality.

Milking practices demand attention to hygiene, precision, and routine. A consistent milking schedule helps stimulate milk production and ensures the goats are comfortable. Proper sanitation of udders and milking equipment minimizes contamination risk and ensures high-quality milk production. Milking can be done by hand or with the assistance of milking machines, and the choice depends on the scale of the operation and the available resources. Proper training in milking techniques is essential for both the goats' efficiency and well-being.

Herd management involves monitoring the dairy goat herd's health, reproduction, and overall performance. Routine health checks, vaccinations, and deworming programs are essential to preventive care. Identifying signs of illness, such as changes in behavior or appetite, allows for prompt intervention and veterinary care.

Breeding practices are carefully managed to optimize the herd's reproductive efficiency, ensuring a consistent milk supply. Monitoring individual milk production, tracking breeding cycles, and maintaining accurate records contribute to effective herd management.

Dairy goat farming also provides opportunities for value-added products beyond fresh milk. Goat cheese, yogurt, and soap are among the products that can be crafted from goat milk, adding potential income streams to the enterprise. Farmers can explore niche markets and direct-to-consumer sales, capitalizing on the increasing demand for artisanal and locally produced goat milk products. Diversifying the product line allows farmers to maximize the economic potential of their dairy goat operation and cater to various consumer preferences.

Engaging in dairy goat farming requires a commitment to continuous learning and staying informed about best practices. Farmers can benefit from joining local or online goat associations, participating in workshops, and networking with experienced farmers. These resources provide valuable insights, support, and opportunities to exchange knowledge with other individuals passionate about dairy goat farming. Staying informed about advancements in goat farming, nutrition, and health care practices contributes to the overall success and sustainability of the dairy goat enterprise.

In conclusion, exploring dairy goat farming offers a pathway to a fulfilling and potentially lucrative venture. From selecting the right breed and providing appropriate housing to implementing sound nutrition and milking practices, successful dairy goat farming requires a holistic and well-informed approach. Dedication to the herd's well-being and a commitment to ongoing learning and adaptation positions farmers for success in this dynamic and rewarding industry. Whether for personal enjoyment, sustainable living, or as a commercial endeavor, dairy goat farming promises a rich and diverse agricultural experience.

Meat Goat Farming: A Comprehensive Guide

Meat goat farming, a dynamic and burgeoning sector within the livestock industry, offers a comprehensive guide for individuals seeking a rewarding venture in raising goats for meat production. Boasting traits such as adaptability, hardiness, and efficient feed conversion, meat goats have become increasingly popular among farmers and homesteaders. Engaging in meat goat farming involves a multifaceted approach encompassing breed selection, housing, nutrition, breeding practices, health management, and overall herd care. This comprehensive guide serves as a roadmap for those venturing into the meat goat industry, providing essential insights into critical aspects that contribute to successful and sustainable meat goat farming.

Breed selection is a pivotal starting point in establishing a thriving meat goat herd. Numerous goat breeds excel in meat production, each with distinct characteristics suitable for specific farming goals. Boer goats, renowned for their rapid growth rates and high-quality meat, have gained widespread popularity. Other breeds, such as the Spanish, Kiko, and Savanna, offer unique attributes that cater to various environmental conditions and breeding objectives. The selection process involves considering factors like adaptability, growth rates, reproductive efficiency, and the intended scale of the meat goat enterprise.

Providing suitable housing and shelter is imperative to ensure the well-being and productivity of meat goats. Goat barns or shelters should offer protection from adverse weather conditions while maintaining proper ventilation. Adequate space per goat, appropriate bedding, and effective waste management contribute to a clean and comfortable environment. Well-designed feeding and watering facilities further enhance the overall efficiency of the operation. Ensuring that the housing infrastructure aligns with the specific needs of meat goats promotes a stress-free and conducive environment for optimal growth and development.

Nutrition plays a crucial role in the success of a meat-goat farming operation. A well-balanced diet tailored to the nutritional needs of growing and mature goats is essential for efficient meat production. High-quality forage, such as clover or Bermuda grass, serves as a primary component of the diet, complemented by grains, minerals, and vitamins. Access to clean water is paramount for hydration and optimal growth. Balancing the nutritional requirements according to the meat goats' age, weight, and production goals ensures their overall health and promotes robust muscle development.

Breeding practices in meat goat farming are carefully managed to optimize reproductive efficiency and the growth potential of the herd. Strategic breeding involves selecting superior genetics, focusing on weight gain, conformation, and resistance to common diseases. Controlled breeding seasons, often synchronized with favorable environmental conditions, contribute to the uniformity and marketability of the offspring. Proper management of kidding, including attentive care during the birthing process, supports the health and vitality of the newborns. Sound breeding practices are crucial to maintaining a sustainable and genetically robust meat goat herd.

Routine health management is a critical component of meat goat farming to prevent diseases and ensure the overall well-being of the herd. Vaccination programs, tailored to the specific needs of the goats and guided by veterinary advice, play a pivotal role in preventing common diseases. Regular health checks, vigilant monitoring for signs of illness, and prompt intervention contribute to maintaining a healthy herd. Effective parasite control, including strategic deworming, aids in preventing internal and external parasites that can impact the growth and productivity of meat goats. A proactive approach to health management is fundamental to a meat-goat farming enterprise's long-term success and sustainability.

Optimizing meat production involves managing the herd for efficient weight gain and carcass quality. Implementing controlled feeding practices, with attention to nutrition and growth rates, contributes to achieving desirable carcass characteristics. Monitoring body condition scores and adjusting feeding regimes as needed ensures that meat goats reach market weight efficiently. Additionally, considering pasture management, rotational grazing, and environmental stressors contributes to overall herd health and productivity.

Marketing strategies play a pivotal role in the success of a meat-goat farming enterprise. They establish a target market, whether local consumers, regional markets, or niche markets, influencing various aspects of the operation, including breeding decisions, herd size, and production methods. Developing strong marketing channels, such as direct-to-consumer sales, partnerships with local markets, or participation in agricultural events, enhances the visibility and profitability of the meat goat enterprise. Building a brand that emphasizes the goats' quality, sustainability, and humane treatment can differentiate the product and attract discerning consumers.

Engaging with the broader community and industry networks provides valuable support and opportunities for meat goat-farmers. Joining local or online goat associations, attending workshops, and participating in agricultural events fosters knowledge-sharing and networking. Collaborating with experienced farmers, veterinarians, and industry experts enables access to insights and resources that contribute to improving meat-goat farming practices. Staying informed about market trends, consumer preferences, and advancements in breeding and management practices ensures that meat goat farmers remain adaptable and responsive to evolving demands.

Meat goat farming offers the potential for diversification through value-added products. Beyond the primary focus on meat production, goat by-products such as hides, milk, and fiber can contribute to additional income streams. Exploring opportunities for value addition, such as processing goat meat into sausages or specialty cuts, allows farmers to maximize the economic potential of their meat goat operation. Diversifying the product line aligns with evolving consumer preferences and market demands, creating a more resilient and adaptable enterprise.

In conclusion, meat-goat farming represents a dynamic and rewarding venture with the potential for economic success and sustainability. From selecting the right breed and providing appropriate housing to implementing sound nutrition, breeding practices, and health management, successful meat goat farming demands a holistic and well-informed approach. The commitment to the herd's well-being and a strategic approach to marketing and continuous learning position farmers for success in this dynamic and expanding sector. Whether for personal consumption, regional markets, or commercial endeavors, meat-goat farming holds the promise of a thriving and resilient agricultural enterprise.

Harvesting Fiber: Angora and Cashmere Goats

Harvesting fiber from Angora and Cashmere goats introduces a fascinating dimension to goat farming, offering unique opportunities for those interested in the luxurious world of natural fibers. Angora goats are renowned for their mohair, a lustrous and silky fiber, while Cashmere goats produce exceptionally soft and insulating cashmere wool. Engaging in the fiber industry involves careful husbandry practices, proper goat herd management, and meticulous fiber harvesting techniques. This comprehensive guide delves into the critical aspects of harvesting fiber from Angora and Cashmere goats, providing insights into breed selection, housing, nutrition, health management, and the intricate process of fiber collection.

Breed selection is a crucial first step in establishing a successful fiber harvesting venture. Angora goats, prized for their long, curly mohair fibers, are available in various breeds, with the Angora and Pygora being popular. On the other hand, cashmere goats include breeds such as the Cashmere, Pashmina, and Nigora. The choice of breed depends on specific fiber characteristics desired, environmental conditions, and the intended scale of the fiber production enterprise. Selecting goats with superior genetics for fiber production ensures the quality and yield of the harvested fibers.

Providing suitable housing and shelter is essential to ensure the well-being and productivity of Angora and Cashmere goats. Goat barns or shelters should be well-ventilated, protecting the goats from extreme weather conditions and maintaining optimal conditions for fiber growth. Adequate space per goat, proper bedding, and effective waste management contribute to a clean and comfortable environment. Well-designed sharing facilities, with secure restraint options, are crucial for efficiently harvesting mohair and cashmere, ensuring the safety of both goats and handlers.

Nutrition plays a pivotal role in the fiber production of Angora and Cashmere goats. A well-balanced diet, tailored to meet the specific nutritional needs of goats producing mohair and cashmere, is essential for fiber growth and overall health. High-quality forage, supplemented with grains, minerals, and vitamins, contributes to developing solid and lustrous fibers. Adequate protein intake is crucial for mohair production, while the fiber-boosting mineral selenium is vital for both Angora and Cashmere goats. Ensuring proper nutrition throughout the year supports the growth and quality of the harvested fibers.

Health management is a critical aspect of fiber harvesting, as the overall well-being of Angora and Cashmere goats directly impacts the quality of their fibers. Regular health checks, vaccination programs, and parasite control are essential components of preventive care. Maintaining a consistent grooming routine, including hoof trimming and coat care, minimizes the risk of matting and fiber contamination. Observing goats for signs of stress or discomfort allows for prompt intervention and veterinary care, ensuring that the fiber production process remains humane and ethical.

Harvesting mohair from Angora goats involves shearing, a delicate and skilled practice. Shearing is typically done twice a year to collect the mohair fibers without causing stress or harm to the goats. Experienced shearers use specialized equipment to carefully remove the mohair fleece, preserving its length and quality. Proper handling and restraint techniques are crucial during shearing to ensure the goats' and handlers' safety and well-being. The harvested mohair is sorted, cleaned, and processed to create a wide range of products, from luxurious textiles to specialty yarns.

Cashmere harvesting from Cashmere goats is a meticulous process that revolves around the natural shedding of the winter undercoat. Cashmere goats naturally shed their fine, downy fibers as spring approaches, offering a suitable harvest time. To collect the cashmere, the goats are gently combed or brushed to remove the loose fibers. This process, known as de-hairing, separates the soft cashmere from the coarser guard hairs, yielding the highly sought-after fiber. Following de-hairing, the cashmere fibers undergo cleaning, sorting, and processing to create premium-quality textiles, garments, and accessories.

Marketing and selling Angora mohair and Cashmere fibers involves understanding the demands of the luxury fiber market. Establishing a target market, which may include artisanal textile producers, yarn enthusiasts, or high-end fashion designers, influences various aspects of the operation. Developing strong marketing channels, such as direct sales, participation in fiber festivals, or collaboration with specialty retailers, enhances the visibility and profitability of the fiber production enterprise. Emphasizing the unique qualities and ethical practices behind Angora mohair and Cashmere fiber production contributes to building a brand that appeals to discerning consumers.

Engaging with the broader fiber community, including

fiber associations, artisanal textile groups, and specialty retailers, provides valuable networking opportunities and support for Angora and Cashmere fiber producers. Participation in fiber festivals, workshops, and industry events fosters knowledge-sharing and facilitates collaboration with like-minded individuals passionate about natural fibers. Staying informed about trends in sustainable and ethical fiber production allows Angora and Cashmere goat farmers to adapt their practices to meet evolving consumer preferences, contributing to the long-term success of the fiber production enterprise.

In conclusion, harvesting fiber from Angora and

Cashmere goats offers a unique and rewarding avenue within goat husbandry. From selecting the right breed and providing appropriate housing to implementing sound nutrition, health management, and the delicate process of fiber harvesting, successful fiber production demands a holistic and well-informed approach. The commitment to humane and ethical practices, a strategic marketing approach, and continuous learning position fiber producers for success in this niche and high-value sector. Angora and Cashmere fiber farming promises a rich and distinctive agricultural experience, whether for personal enjoyment, artisanal crafting, or commercial endeavors.

CHAPTER X

Marketing and Selling Your Products

Building Your Brand as a Goat Guru

Building your brand as a Goat Guru involves a strategic and multifaceted approach to establishing a reputation as a knowledgeable and trusted authority in goat farming. Whether you are a seasoned goat farmer, an industry expert, or an aspiring author, crafting a distinctive brand is essential for connecting with your target audience, fostering trust, and positioning yourself as a go-to resource in the goat community.

The foundation of building your brand starts with a transparent and authentic identity. Define your mission, values, and unique selling propositions that set you apart in the goat industry. Consider what aspects of goat husbandry you are most passionate about, whether it's sustainable farming practices, ethical treatment of animals, or innovative goat management techniques. Your brand identity should reflect your genuine commitment to the well-being of goats and your expertise in the field.

Consistent and compelling storytelling plays a pivotal role in shaping your brand narrative. Share your journey in goat farming, detailing the challenges you've overcome, lessons learned, and milestones achieved. Authentic storytelling connects with your audience, allowing them to relate to your experiences and gain insights into your expertise. Whether through blog posts, social media updates, or written publications, storytelling humanizes your brand and establishes a relatable persona as a Goat Guru.

Engaging with your audience through various channels is crucial for building a solid brand presence. Leverage social media platforms, forums, and industry events to connect with fellow goat enthusiasts, farmers, and potential clients. Consistent and meaningful interactions build a community around your brand, fostering trust and loyalty. Share valuable content, participate in discussions, and showcase your expertise to position yourself as a credible source of information in the goat community.

Educational content creation is a cornerstone of establishing yourself as a Goat Guru. Develop a content strategy that aligns with your brand identity and addresses the needs of your target audience. Whether through written articles, video tutorials, or podcasts, share practical insights, tips, and advice on various aspects of goat farming. Position yourself as a go-to resource for novice and experienced goat enthusiasts, offering valuable and actionable information that enhances their knowledge and skills.

Authenticity and transparency are integral to building trust with your audience. Share your successes and challenges openly, demonstrating a genuine commitment to the well-being of goats and ethical farming practices. When your audience sees the person behind the brand as relatable and trustworthy, it strengthens the connection and fosters a positive perception of your expertise.

Creating a visually appealing and cohesive brand aesthetic enhances your professional image as a Goat Guru. Invest in a well-designed logo, consistent color schemes, and visually appealing graphics that resonate with your brand identity. A polished and cohesive visual presence across your website, social media, and promotional materials contributes to a professional and memorable brand image.

Networking within the goat community and broader agricultural industry is instrumental in building your brand as a Goat Guru. Attend industry events, collaborate with other goat enthusiasts, and seek opportunities to share your expertise. Building relationships with fellow farmers, industry experts, and potential clients expands your network and enhances the credibility of your brand within the goat community.

Public speaking engagements, workshops, and webinars provide platforms to showcase your expertise and elevate your status as a Goat Guru. Consider offering educational sessions at agricultural conferences, local community events, or online platforms. Sharing your knowledge in a live setting positions you as an authority and allows you to directly connect with your audience, answering their questions and addressing their concerns.

Consistent branding across all touchpoints reinforces your identity as a Goat Guru. Ensure that your messaging, visual elements, and overall brand image align seamlessly across your website, social media profiles, business cards, and promotional materials. Consistency fosters recognition and reinforces your brand's trustworthiness in your audience's eyes.

Testimonials and endorsements from satisfied clients, collaborators, or industry peers add credibility to your brand as a Goat Guru. Encourage and showcase positive feedback on your website, social media, and promotional materials. Real-life testimonials provide evidence of the value you bring to the goat community and contribute to the overall trustworthiness of your brand.

Building a brand as a Goat Guru requires a long-term commitment to excellence, continuous learning, and a genuine passion for goat farming. By cultivating a distinctive identity, sharing your knowledge through various channels, engaging with your audience authentically, and maintaining consistency across all touchpoints, you can establish a reputable brand that resonates with goat enthusiasts, farmers, and industry stakeholders. As you navigate the exciting journey of building your brand, remember that authenticity, expertise, and a genuine love for goats will be the cornerstones of your success as a Goat Guru.

Marketing Goat Products Locally and Online

Whether locally or online, marketing goat products demands a thoughtful and strategic approach to reach a diverse audience of consumers interested in quality goat-derived goods. Local and online marketing efforts provide distinct opportunities to connect with consumers, and combining these strategies can maximize the visibility and profitability of your goat product business.

Local marketing involves engaging with the community around your farm or business. Establishing a solid local presence begins with understanding the needs and preferences of the local consumer base. Attend local farmers' markets, fairs, and community events to showcase your goat products. Create eye-catching displays that highlight the uniqueness and quality of your offerings, fostering a direct connection with potential customers. Networking with other local businesses, such as restaurants, grocery stores, or artisanal shops, can open avenues for collaborations and expanded distribution channels.

Utilize traditional marketing methods to promote your goat products locally. Distribute flyers, posters, or business cards at local community centers, grocery stores, and bulletin boards. Advertise in local newspapers or community newsletters to reach a broader audience. Consider organizing workshops or events on your farm, offering hands-on experiences or educational sessions related to goat farming. Building a local customer base through personal interactions and community engagement establishes trust and loyalty, which are vital for a successful goat product business.

In online marketing, a well-designed website serves as the virtual storefront for your goat products. Ensure that your website is user-friendly, visually appealing, and provides comprehensive information about your offerings. Feature high-quality images and detailed descriptions of each product, highlighting the craftsmanship, quality, and uniqueness that set your goat products apart. Implement secure and convenient online purchasing options, making it easy for customers to buy your products directly from your website.

Social media platforms are pivotal in online marketing, offering a dynamic space to showcase your goat products and connect with a broader audience. Utilize platforms such as Instagram, Facebook, and Pinterest to share visually engaging content, including images, videos, and stories about your farm, goats, and product creation process. Engage with your audience by responding to comments, conducting polls, or sharing behind-the-scenes glimpses of your goat farming journey. Social media provides a powerful tool to build a community around your brand and create a loyal customer base.

Online marketplaces and e-commerce platforms offer additional avenues to expand your reach. List your goat products on popular online marketplaces like Etsy, Amazon, or local e-commerce platforms. This allows you to tap into a global market and attract customers needing local access to your products. Leverage the convenience and accessibility of online platforms to showcase the versatility and appeal of your goat products to a diverse and geographically dispersed audience.

Search engine optimization (SEO) ensures your goat products are discoverable online. Optimize your website and online product listings with relevant keywords, meta tags, and descriptions to enhance visibility in search engine results. Implementing SEO strategies helps potential customers find your goat products when they search for related terms, driving organic traffic to your website and online listings.

Content marketing is a valuable online marketing strategy that involves creating and sharing relevant content to attract and engage your target audience. Develop blog posts, articles, or guides related to goat husbandry, sustainable farming practices, or the benefits of goat products. Share this content on your website and social media platforms to position yourself as an authority in the goat industry. Educational and informative content attracts potential customers and establishes your brand as a trusted source of knowledge.

Email marketing remains an effective tool for nurturing relationships with existing customers and reaching potential buyers. Build an email list by encouraging website visitors and customers to subscribe for updates, promotions, or newsletters. Regularly communicate with your email subscribers, sharing updates about new products, promotions, or upcoming events. Personalized and targeted email campaigns help retain customer loyalty and encourage repeat purchases.

Customer reviews and testimonials are influential in online marketing. Encourage satisfied customers to leave reviews on your website, social media pages, or online marketplaces. Positive reviews build credibility and trust, influencing potential customers to choose your goat products. Responding to customer reviews, whether positive or addressing concerns, demonstrates transparency and a commitment to customer satisfaction, further enhancing your online reputation.

Combining local and online marketing efforts creates a holistic approach to promoting your goat products. Cross-promote your online presence in local events and vice versa. Share stories and updates about local community engagement on your website and social media platforms. Encourage local customers to follow your online channels for the latest updates and promotions. This integrated strategy reinforces your brand identity and creates a seamless experience for customers, whether they encounter your goat products in a local market or on your website.

In conclusion, marketing goat products locally and online requires a strategic and multifaceted approach to reach a diverse and engaged audience. Local marketing efforts, including community events, collaborations, and traditional advertising, build a strong foundation within your immediate vicinity. Through a well-optimized website, social media platforms, e-commerce, and content marketing, online marketing extends your reach globally. The synergy between local and online marketing creates a comprehensive strategy that maximizes visibility, fosters customer loyalty, and positions your goat product business for long-term success in a competitive market.

Navigating Regulations and Compliance

Navigating regulations and compliance is crucial to managing a successful goat farming operation. Whether you're a novice herdsman or an experienced goat farmer, understanding and adhering to local, regional, and national regulations is essential for the well-being of your goats, the sustainability of your farm, and compliance with legal standards.

Local regulations often vary, and it's vital to familiarize yourself with the specific requirements in your area. Local ordinances may cover zoning laws, land use regulations, and restrictions on the number of animals allowed on a property. Ensure that your goat farming activities align with these local regulations to prevent potential legal issues and maintain positive relations with neighbors and the community.

Regional and national regulations set animal welfare, health, and disease control standards. Familiarize yourself with the guidelines and requirements established by agricultural agencies and departments responsible for overseeing livestock farming. Compliance with these regulations often involves implementing proper biosecurity measures, maintaining accurate health records, and participating in disease control programs.

Regular communication with local agricultural extension offices and veterinary services can provide valuable insights and assistance navigating these regional and national regulations.

Environmental regulations are another critical aspect to consider in goat farming. Proper waste management, runoff control, and environmental conservation practices are often mandated to minimize the ecological impact of farming operations. Develop and implement strategies for waste disposal, manure management, and pasture rotation that align with environmental regulations. Sustainable farming practices ensure compliance and contribute to the long-term health of your land and surrounding ecosystems.

Food safety authorities regulate quality and safety standards for goat products. If you're involved in producing and selling goat milk, cheese, or meat, it's imperative to adhere to the hygiene and safety guidelines set by these authorities. Implementing good manufacturing practices, maintaining cleanliness in processing areas, and ensuring the health and well-being of your goats contribute to producing safe and high-quality goat products that meet regulatory standards.

Animal welfare regulations govern farm animals' ethical treatment and care, including goats. Ensure that your goat farming practices align with these regulations to provide humane and ethical treatment to your animals. Adequate housing, nutrition, veterinary care, and handling procedures contribute to meeting animal welfare standards. Stay informed about updates and changes to these regulations to continuously improve the well-being of your goats and maintain a positive reputation for your farm.

Biosecurity measures are crucial in disease prevention and compliance with health regulations. Implement practices to control the spread of diseases among goats, including quarantine protocols for new arrivals, vaccination programs, and regular health monitoring. Collaborate with veterinarians to develop and implement biosecurity plans that align with regional and national disease control regulations. Proactive disease prevention measures not only ensure compliance but also safeguard the health of your goat herd.

Record-keeping is a fundamental aspect of regulatory compliance in goat farming. Maintain accurate and up-to-date records of your goat herd, including health records, breeding history, and any treatments administered. Record-keeping is essential for compliance with regulations and facilitates efficient farm management, aids in disease traceability, and supports decision-making processes.

Participating in educational programs and workshops offered by agricultural extension services, veterinary offices, and industry associations can provide valuable insights into regulatory compliance. Stay informed about changes in regulations, emerging best practices, and advancements in goat farming through continuous learning and networking within the agricultural community. Proactively seeking information and guidance ensures that your goat farming operation complies with evolving standards.

Regular inspections and audits may be conducted by relevant authorities to assess regulation compliance. Prepare for these inspections by maintaining a well-organized and clean farm, ensuring that records are readily available, and promptly addressing potential issues. Open communication with regulatory officials during inspections demonstrates your commitment to compliance and a willingness to collaborate for the betterment of your farm and the industry.

In conclusion, navigating regulations and compliance is fundamental to responsible and sustainable goat farming. Local, regional, and national regulations cover various aspects of goat farming, environmental impact, product quality, and animal welfare. Understanding and adhering to these regulations ensure legal compliance and contribute to the overall success and reputation of your goat farming operation. Regular communication with local agricultural authorities, veterinary services, and industry associations, coupled with ongoing education and proactive biosecurity measures, establishes a solid foundation for regulatory compliance in goat farming.

CHAPTER XI

Troubleshooting Common Challenges

Dealing with Aggressive Behavior

Dealing with aggressive behavior in goats is a crucial aspect of goat management, requiring a nuanced understanding of the causes and effective strategies for prevention and resolution. Aggression in goats can manifest in various forms, including headbutting, biting, or aggressive posturing, and it can be directed towards other goats, humans, or even other animals. Addressing aggressive behavior is essential for the goats' and their caretakers' safety and well-being and for maintaining a harmonious and productive herd.

Understanding the root causes of aggression is the first step in addressing this behavior. Aggression in goats can result from various factors, such as hierarchy establishment within the herd, competition for resources, territorial disputes, fear, pain, or improper handling. Observing the specific triggers and context in which aggression occurs helps to identify the underlying causes and tailor appropriate interventions.

Establishing a clear and well-defined hierarchy within the goat herd is a natural behavior that can reduce the incidence of aggression. However, disruptions in the hierarchy, such as the introduction of new goats or changes in herd dynamics, can trigger aggressive behavior as goats vie for dominance. Gradual introductions of new goats can mitigate aggression, allowing them to acclimate and establish their place within the hierarchy. Providing ample space, feeding stations, and other resources helps minimize competition and reduces the likelihood of aggression related to resource disputes.

Proper handling and socialization from an early age play a vital role in preventing aggressive behavior in goats. Goats accustomed to human interaction and handling are less likely to exhibit fear-based aggression. Regular interaction, positive reinforcement, and gentle handling contribute to building trust between goats and their caretakers. Additionally, early socialization with other goats fosters positive relationships within the herd, reducing the likelihood of aggressive encounters.

Pain or discomfort can be a significant trigger for aggressive behavior in goats. Health issues, injuries, or hoof problems may cause goats to become defensive or irritable. Regular health checks, prompt identification and treatment of any medical problems, and providing appropriate veterinary care contribute to preventing aggression stemming from pain or discomfort.

Territorial behavior is inherent in goats, and aggression can arise when goats perceive a threat to their territory or personal space. Providing adequate space, well-designed enclosures, and strategic placement of resources, such as feeding and watering stations, helps minimize territorial disputes. Overcrowding can contribute to stress and competition, leading to aggressive behavior, so ensuring that the living environment meets the goats' spatial needs is crucial.

Effective management strategies play a pivotal role in dealing with aggressive behavior in goats. Separating aggressive individuals from the herd temporarily can help diffuse tension and allow for individual assessments of the root causes of aggression. Utilizing physical barriers or introducing partitions within the living space can prevent confrontations and reduce the risk of injury. Identifying and removing high-risk items, such as feed buckets or structures perceived as competition triggers, can also contribute to a more peaceful herd dynamic.

Training and desensitization techniques can be employed to modify aggressive behavior in goats. Positive reinforcement, such as rewarding calm behavior, encourages goats to associate positive experiences with specific actions. Consistent and patient training, using verbal commands and gentle handling, reinforces desired behaviors and helps goats understand appropriate responses in various situations. Redirecting aggressive behavior towards positive activities, such as engaging in enrichment activities or following commands, provides an outlet for their instincts in a controlled manner.

When addressing aggressive behavior, it's crucial to prioritize the safety of both goats and handlers. Protective gear, such as gloves and sturdy footwear, should be worn when working with goats exhibiting aggressive tendencies. Maintaining a calm and assertive demeanor during interactions helps establish authority without inciting fear. Avoiding sudden movements or loud noises, which can trigger aggressive responses, contributes to a more relaxed and controlled environment.

In some cases, hormonal factors may contribute to aggressive behavior, particularly in intact males. Castration is a common practice to mitigate aggressive tendencies in male goats, reducing territorial behavior and minimizing the drive to establish dominance. Consulting with a veterinarian to determine the most appropriate timing and method for castration is essential to ensure the well-being of the goat and prevent unintended consequences.

Understanding the individual personalities and dynamics within the herd is crucial for effective aggression management. While specific preventive measures and interventions can be applied broadly, recognizing each goat's unique traits and triggers allows for tailored strategies. Regular observation, documentation of behavioral patterns, and swift intervention, when signs of aggression emerge, contribute to a proactive approach to maintaining a peaceful and cooperative goat herd.

In conclusion, dealing with aggressive behavior in goats requires a multifaceted approach that combines preventive measures, effective management strategies, and a deep understanding of goat behavior. Addressing the root causes, whether related to hierarchy establishment, competition for resources, fear, pain, or improper handling, allows for targeted interventions. Implementing proactive measures, such as proper socialization, training, and health care, creates a positive and harmonious environment for goats. Caretakers can successfully navigate and mitigate aggressive behavior by prioritizing safety, employing management strategies, and tailoring interventions to individual goats, fostering a thriving and content goat herd.

Solving Dietary Issues

Solving dietary issues in goats is a fundamental aspect of effective herd management, as a well-balanced and nutritionally sound diet is paramount to their overall health, productivity, and well-being. Dietary issues can manifest in various forms, ranging from nutritional deficiencies to overconsumption of certain elements, and addressing these concerns requires a comprehensive understanding of goat nutritional requirements, forage quality, and potential challenges that may arise in their diet.

The foundation of solving dietary issues lies in a thorough understanding of the nutritional needs of goats at different life stages and production levels. Goats, being ruminants, have unique digestive systems that require a mix of fiber, energy, protein, vitamins, and minerals for optimal health. A lack of essential nutrients or an imbalance in their ratios can lead to various health issues, including poor growth, reproductive problems, and compromised immune function.

Forage quality plays a pivotal role in addressing dietary issues in goats, as forages often constitute a significant portion of their diet. Ensuring access to high-quality forages that meet their nutritional requirements is essential. Regular analysis of forage samples can provide valuable insights into their nutrient content, allowing caretakers to adjust diets and supplements where necessary. Understanding seasonal variations in forage quality is also crucial, as nutrient levels may fluctuate based on weather, soil conditions, and plant maturity.

Supplementing the goat diet strategically is a critical component of solving dietary issues. While forages form the basis of their nutrition, supplemental feed may be required to meet specific needs, especially when forage quality is deficient. Protein, energy, and mineral supplements can be introduced to address nutritional gaps, but it's essential to tailor supplementation to the specific requirements of the herd. Over-supplementation can lead to imbalances and pose health risks, emphasizing the importance of precision in dietary management.

Addressing dietary issues may also involve managing access to certain plants or feedstuffs that can be toxic or harmful to goats. Goats are notorious for their browsing habits, and while their selective feeding behavior can be advantageous in foraging, it also puts them at risk of consuming toxic plants. Familiarizing oneself with the local flora and removing or restricting access to potentially harmful plants is a preventive measure to mitigate toxicity-related dietary issues.

Water quality and availability are often overlooked but critical factors in solving dietary issues. Clean, fresh water is essential for digestion, nutrient absorption, and overall hydration. Inadequate water intake can lead to various health problems, including dehydration and urinary calculi. Regular monitoring of water sources, especially in arid or extreme weather conditions, ensures that goats can access an adequate and clean water supply.

Another common dietary issue in goats is the development of metabolic disorders, such as ruminal acidosis or ketosis. These conditions can arise from imbalances in the diet, mainly when there is a rapid change in the type or quantity of feed. Gradual transitions between diets, providing a consistent and well-balanced ratio, and avoiding sudden shifts in feed sources are effective strategies for preventing metabolic disorders in goats.

Maintaining a healthy body condition score is a practical way to assess and manage dietary issues in goats. Body condition scoring involves evaluating the amount of fat covered over specific areas of the goat's body, providing insights into their nutritional status. Monitoring body condition regularly allows caretakers to make informed decisions regarding dietary adjustments, ensuring that goats maintain an optimal weight and are not undernourished or over-conditioned.

The management of parasites is closely tied to dietary issues, as internal parasites can contribute to nutritional deficiencies in goats. A robust parasite control program involving regular deworming, pasture management, and monitoring fecal egg counts helps prevent parasitic infestations that can compromise the herd's nutritional status. Integrating sustainable and holistic approaches to parasite control, such as rotational grazing and herbal dewormers, aligns with a comprehensive dietary management strategy.

In the context of solving dietary issues, caretakers should also be attuned to the specific needs of different goat breeds and variations based on age, sex, and production purpose. For example, dairy goats may have higher energy and protein requirements during lactation, while growing kids necessitate diets that support proper development. Tailoring diets to meet the specific requirements of different groups within the herd contributes to optimal health and performance.

Regular monitoring of the herd's overall health and individual animals' performance is essential in promptly identifying and addressing dietary issues. Signs of poor growth, coat condition, reproductive problems, or behavioral changes can indicate nutritional deficiencies or imbalances. Implementing routine health checks, including body condition scoring, weight assessments, and observation of feeding behavior, allows caretakers to detect dietary issues early and intervene effectively.

Consulting with a veterinarian or nutritionist can provide valuable insights and guidance in addressing complex dietary issues. A professional assessment of the herd's nutritional needs, analysis of forage quality, and formulation of customized diets can enhance the effectiveness of dietary management. Veterinarians can also conduct blood tests or other diagnostic evaluations to identify specific nutritional deficiencies or imbalances, guiding targeted interventions.

In conclusion, solving dietary issues in goats requires a comprehensive and proactive approach that considers their unique nutritional requirements, forage quality, and potential challenges. A well-balanced diet, strategic supplementation, access to clean water, and preventive measures against toxicity and metabolic disorders contribute to maintaining the health and productivity of the goat herd. Regular monitoring, early intervention, and collaboration with veterinary professionals ensure that dietary issues are identified and addressed promptly, fostering a thriving and resilient goat farming operation.

Addressing Common Goat Ailments

Addressing common goat ailments is an integral aspect of responsible goat husbandry, requiring a vigilant and proactive approach to ensure the health and well-being of the herd. Like all animals, goats are susceptible to various illnesses, and caretakers must be well-versed in recognizing symptoms, implementing preventive measures, and administering appropriate treatments to maintain a thriving and resilient herd.

One prevalent health concern in goats is parasitic infestations, particularly internal parasites such as gastrointestinal worms. These parasites can cause various issues, including weight loss, lethargy, and anemia. Implementing a robust parasite control program involving regular deworming, pasture management, and monitoring fecal egg counts is essential. Utilizing a variety of deworming agents and adopting sustainable practices, such as rotational grazing and herbal dewormers, helps prevent the development of resistance and maintains the efficacy of parasite control measures.

Foot rot is another common ailment in goats,

characterized by inflammation and infection of the hoof tissues. This condition is often exacerbated in damp or muddy conditions. Maintaining clean and dry living environments, regular hoof trimming, and prompt treatment of any signs of foot rot, such as lameness or foul odor, are crucial preventive measures. Topical antiseptics and antibiotics may be administered under veterinary guidance to address active cases of foot rot and prevent its spread within the herd.

Respiratory infections like pneumonia can affect goats,

especially in crowded or poorly ventilated environments. Stress, sudden changes in weather, or exposure to drafts can contribute to respiratory issues. Preventive measures include ensuring proper ventilation in housing, avoiding overcrowding, and minimizing stress factors. Vaccination against common respiratory pathogens, such as Mannheimia haemolytica and Pasteurella multocida, can be a proactive strategy to protect the herd.

Clostridial diseases, including enterotoxemia (overeating disease) and tetanus, pose a significant threat to goats. Bacteria of the Clostridium genus cause these diseases, and preventive measures involve timely vaccination. Vaccination against clostridial diseases should be administered according to a recommended schedule, with booster shots to maintain immunity. Also, proper management practices, such as gradual diet changes and wound care to prevent tetanus, reduce the risk of these diseases.

Caprine arthritis encephalitis (CAE) is a viral infection that affects goats, causing arthritis, mastitis, and neurological symptoms. Preventing the spread of CAE involves testing and segregating seropositive goats, as the virus is primarily transmitted through colostrum and milk. Implementing strict biosecurity measures, such as using separate feeding equipment for kids and adults, helps prevent the transmission of CAE within the herd. Testing and culling seropositive animals are essential to a comprehensive CAE control program.

Goats are also susceptible to external parasites, such as lice and mites, which can cause itching, hair loss, and skin irritation. Regular grooming, especially during the winter months when these parasites are more prevalent, helps detect and address external infestations promptly. Topical treatments, such as insecticidal sprays or dust, can eliminate external parasites. Additionally, providing clean bedding and maintaining hygienic living conditions contribute to preventing external parasite infestations.

Mastitis, an inflammation of the mammary gland, is a concern in dairy goats and can impact milk production and quality. Practicing proper hygiene during milking, promptly addressing any signs of mastitis, such as swelling or changes in milk quality, and ensuring appropriate nutrition contribute to preventing and managing mastitis. Culling goats with chronic mastitis and implementing a mastitis prevention program are crucial to maintaining udder health in dairy herds.

While these are common goat ailments, individual goats may also experience various other health issues, such as urinary calculi, urinary tract infections, or metabolic disorders. A thorough understanding of goat anatomy, behavior, and joint health concerns is essential for caretakers to recognize signs of illness promptly. Regular health checks, observation of behavioral changes, and timely veterinary consultations enhance the caretaker's ability to address a wide array of health issues in goats.

Timely and accurate diagnosis is pivotal in addressing goat ailments effectively. Establishing a solid relationship with a veterinarian specializing in small ruminant health is an invaluable resource for goat caretakers. Veterinary professionals can conduct diagnostic tests, such as bloodwork, fecal examinations, or bacterial cultures, to identify the specific pathogens or causes of illness. In cases where a definitive diagnosis may be challenging, consulting with experienced veterinarians or specialists in goat health ensures that the appropriate treatment protocols are implemented.

Caretakers should be familiar with basic first aid principles and have a well-equipped kit readily available. Providing immediate care for minor injuries, such as cuts or scrapes, and addressing fundamental health concerns, such as administering oral medications or wound cleaning, allows caretakers to respond promptly to emergent situations. Training in basic veterinary procedures under the guidance of a veterinarian empowers caretakers to provide initial care and support until professional veterinary assistance is available.

Preventive measures, including vaccination, proper nutrition, and biosecurity practices, form the foundation for addressing common goat ailments. Vaccination protocols should be developed in consultation with a veterinarian based on the specific health risks in the geographical region and the herd's unique characteristics. Maintaining a well-balanced and nutrient-dense diet tailored to the particular needs of different groups within the herd contributes to overall health and resilience.

Biosecurity practices are essential in preventing the introduction and spread of infectious diseases within the goat herd. Implementing quarantine protocols for new arrivals, restricting access to visitors, and practicing proper sanitation minimize the risk of disease transmission. Biosecurity measures should extend to equipment, clothing, and other items that may come in contact with the goats to prevent the accidental spread of pathogens.

Regular monitoring and record-keeping are crucial components of managing goat health. Keeping detailed records of health checks, vaccinations, treatments, and any observed behavioral changes aids in identifying patterns and trends that may indicate potential health issues. Monitoring body condition scores, weight changes, and reproductive performance allows caretakers to detect health concerns early and implement interventions.

In conclusion, addressing common goat ailments requires a holistic and proactive approach encompassing preventive measures, prompt diagnosis, and effective treatment strategies. A deep understanding of goat physiology, behavior, and joint health concerns is essential for caretakers to provide optimal care. Collaboration with veterinary professionals, ongoing education, and a commitment to biosecurity practices contribute to maintaining a healthy and thriving goat herd. Regular health checks, timely interventions, and a responsive approach to emerging health issues position caretakers as stewards of the well-being of their goats, ensuring a sustainable and resilient goat farming operation.

CHAPTER XI

. Future Trends in Goat Farming

Sustainable Practices for Goat Herding

Implementing sustainable practices in goat herding is paramount to ensuring the long-term health and productivity of the herd while also promoting environmental conservation and ethical stewardship. Sustainable goat herding involves a holistic approach that considers the well-being of the goats, the impact on the surrounding ecosystem, and the economic viability of the farming operation. From pasture management to waste reduction and ethical treatment of animals, caretakers can adopt a range of practices to create a harmonious and sustainable goat herding enterprise.

Pasture management is a cornerstone of sustainable goat herding, emphasizing rotational grazing and thoughtful land use. By periodically dividing pasture into smaller paddocks and rotating goats, caretakers prevent overgrazing, promote optimal forage growth, and minimize soil erosion. This approach maintains pasture health and ensures that goats have access to diverse forage, meeting their nutritional needs. Integrating legumes and other nitrogen-fixing plants into pastures contributes to soil fertility, reducing the reliance on synthetic fertilizers and promoting a self-sustaining ecosystem.

Waste management is a critical aspect of sustainability in goat herding. Manure generated by goats is a valuable resource when appropriately managed. Implementing composting systems turns goat manure into nutrient-rich organic fertilizer that can enhance soil fertility.

Composted manure improves soil structure, retains moisture, and provides essential nutrients to plants, contributing to sustainable pasture management. By recycling manure on-site, caretakers reduce the environmental impact of waste disposal while enriching the farm's soil health.

Ethical and humane treatment of goats is central to sustainable herding practices. Providing appropriate housing, adequate space, and access to clean water ensures the herd's well-being. Goat-friendly handling methods that minimize stress during activities such as milking, hoof trimming, and medical treatments contribute to a positive and respectful relationship between caretakers and goats. Sustainable goat herding also involves addressing goats' social and behavioral needs and encouraging natural behaviors such as browsing and social interaction.

Implementing integrated pest management (IPM) strategies is a sustainable approach to controlling parasites and diseases in goats. Rather than relying solely on chemical interventions, caretakers can integrate biological control methods, such as beneficial insects or rotational grazing with other livestock species, to manage parasite populations. This reduces the risk of developing dewormer resistance and promotes a balanced and resilient ecosystem. Regular monitoring of goat health, coupled with targeted treatments when necessary, helps maintain a healthy herd without relying solely on chemical interventions.

Water conservation is critical in sustainable goat herding, particularly in regions prone to drought or water scarcity. Implementing water-efficient systems, such as drip irrigation in pastures and rainwater harvesting, reduces the reliance on traditional water sources. Adequate hydration is essential for goat health, and sustainable water management practices benefit the farm and contribute to overall water conservation in the broader context of agriculture.

Selecting appropriate goat breeds for the local environment is a sustainable practice that enhances the resilience of the herd. Local or heritage breeds adapted to the climate and forage conditions are often more robust and better suited to the region's challenges. These breeds may also have natural resistance to certain diseases, reducing the need for intensive medical interventions. By supporting genetic diversity and preserving local breeds, caretakers contribute to the sustainability of goat herding practices and help maintain the unique characteristics of regional goat populations.

Agroforestry practices, such as integrating trees and shrubs into pasture areas, contribute to sustainable goat herding by providing additional forage, shade, and environmental benefits. Trees can offer shelter to goats, reduce heat stress, and improve overall pasture productivity. Additionally, agroforestry systems enhance biodiversity, attract beneficial insects, and contribute to carbon sequestration. Integrating diverse vegetation types also creates a more resilient and dynamic ecosystem that supports the well-being of goats and the surrounding environment.

Energy efficiency is a crucial consideration in sustainable goat herding, encompassing the use of renewable energy sources and minimizing the carbon footprint of the farming operation. Solar or wind-powered systems for water pumping, lighting, and other energy needs contribute to a more sustainable and environmentally friendly operation. Implementing energy-efficient practices, such as optimizing building design for natural ventilation and utilizing energy-efficient appliances, reduces reliance on non-renewable energy sources and aligns with the principles of sustainable goat herding.

Community engagement and collaboration are integral components of sustainable goat herding practices. Building connections with local communities, sharing knowledge, and participating in cooperative initiatives contribute to the broader sustainability of the agricultural landscape. Collaborative efforts can include sharing resources, implementing collective pest management strategies, or marketing goat products. Sustainable goat herding becomes a shared responsibility beyond individual farms by fostering a sense of community and cooperation.

Education and outreach play a vital role in promoting sustainable goat herding practices. Sharing knowledge about sustainable farming methods, ethical animal treatment, and the environmental impact of goat herding fosters awareness and encourages others to adopt similar practices. Educational programs, workshops, and partnerships with agricultural extension services contribute to a culture of sustainability within the broader farming community. By disseminating information and best practices, caretakers contribute to improving sustainable goat herding across diverse farming landscapes.

In conclusion, sustainable practices for goat herding encompass a holistic and ethical approach that considers the well-being of the goats, environmental conservation, and economic viability. From pasture management and waste reduction to ethical treatment of animals and community engagement, caretakers can adopt a range of practices to create a harmonious and sustainable goat herding enterprise. By prioritizing sustainable methods, caretakers contribute to the resilience of their own farms and play a crucial role in the broader movement toward sustainable agriculture and responsible stewardship of natural resources.

Emerging Technologies in Goat Farming

Emerging technologies in goat farming are revolutionizing traditional practices, offering innovative solutions to enhance goat herds' efficiency, productivity, and overall management. From precision farming and digital health monitoring to advanced reproductive technologies, these innovations are reshaping the landscape of goat farming, providing caretakers with powerful tools to optimize herd health, streamline operations, and make data-driven decisions.

One significant advancement is the integration of precision farming technologies, which involve using sensors, GPS, and data analytics to monitor and manage various aspects of goat farming. Smart collars equipped with sensors can track individual goats' location and movement patterns, offering insights into grazing behavior, herd dynamics, and overall activity levels. This information enables caretakers to optimize pasture management, identify potential health issues, and ensure that goats access sufficient forage.

Digital health monitoring systems have also emerged as a transformative technology in goat farming. Wearable devices with sensors can track vital signs, such as heart rate, body temperature, and respiratory rate, providing real-time health data for individual goats. This proactive monitoring allows caretakers to detect early signs of illness, implement timely interventions, and reduce the risk of disease spread within the herd. Integrating artificial intelligence (AI) in analyzing health data further enhances the accuracy and efficiency of disease detection.

Reproductive technologies have seen significant advancements, offering tools to optimize breeding programs and genetic selection in goat herds. In vitro fertilization (IVF) and embryo transfer techniques allow for the rapid propagation of superior genetic traits, accelerating the improvement of herd genetics.

Additionally, artificial insemination (AI) enables caretakers to access genetic material from high-performing bucks, expanding the herd's genetic diversity and improving overall productivity.

Precision nutrition is another area where technology is making a substantial impact on goat farming. Automated feeding systems with sensors and algorithms can precisely dispense customized rations based on individual nutritional needs. This optimizes each goat's nutritional intake and reduces feed wastage, contributing to overall operational efficiency. Advanced feed analysis technologies allow caretakers to assess the nutritional content of forages and supplements, helping formulate well-balanced diets that enhance the health and productivity of the herd.

Robotics is transformative in various aspects of goat farming, particularly in automated milking systems. Robotic milking machines offer a hands-free and stress-free milking experience for goats. These systems can recognize individual goats, monitor milk yield, and provide valuable data on milk composition. Automating milking processes reduces labor demands and allows for more frequent and precise milking, contributing to increased milk production and quality.

Data analytics and farm management software are essential for caretakers to make informed decisions and optimize overall herd performance. These platforms integrate data from various sources, including health monitoring systems, reproductive records, and production metrics, providing a comprehensive herd overview. Analyzing this data enables caretakers to identify trends, predict potential issues, and implement targeted interventions to enhance productivity and well-being.

Remote sensing technologies, such as satellite imagery and drones, also find applications in goat farming. These tools can assess pasture health, monitor vegetation density, and identify areas of potential concern.

Drones equipped with thermal imaging cameras can aid in locating lost or sick goats, particularly in extensive grazing systems. Satellite technology contributes to precision farming by providing insights into land use, vegetation quality, and potential environmental impacts.

Biometric technologies, including facial recognition and ear tag systems, offer unique identification methods for individual goats. These systems facilitate accurate record-keeping, pedigree tracking, and monitoring each goat's lifecycle. Biometric data can be integrated into farm management software, streamlining administrative tasks and providing caretakers with a comprehensive understanding of the individual histories and characteristics of each goat in the herd.

Blockchain technology is also making inroads in goat farming, particularly in traceability and transparency in the supply chain. Using blockchain, caretakers can create a secure and immutable record of each goat's journey from farm to market. This enhances the credibility of the goat products and allows consumers to access detailed information about the origin, health, and welfare of the goats they support.

Environmental monitoring systems, including sensors for air quality, temperature, and humidity, contribute to the overall well-being of goats by ensuring optimal living conditions. These systems can alert caretakers to issues like heat stress or inadequate ventilation, allowing for timely interventions. Additionally, they contribute to sustainable farming practices by minimizing environmental impact and promoting resource-efficient operations.

While emerging technologies offer numerous benefits, their successful integration into goat farming requires caretakers to adapt and acquire new skills. Training programs and educational resources play a crucial role in helping caretakers understand the functionalities and applications of these technologies. By fostering a culture of innovation and continuous learning, the goat farming community can harness the full potential of emerging technologies to elevate the standards of care, productivity, and sustainability in goat herding operations.

In conclusion, integrating emerging technologies is transforming the landscape of goat farming, offering unprecedented opportunities to optimize herd management, enhance productivity, and improve overall sustainability. From precision farming and digital health monitoring to advanced reproductive technologies and blockchain traceability, these innovations empower caretakers with powerful tools to make informed decisions and elevate the standards of goat farming practices. As technology advances, the goat farming industry stands at the forefront of a new era, where data-driven insights and automation contribute to the well-being of goats, the efficiency of operations, and the overall success of sustainable goat herding enterprises.

Joining the Goat Community: Events and Networks

Joining the goat community is a dynamic and enriching experience for caretakers, offering opportunities to connect with fellow enthusiasts, share knowledge, and stay abreast of the latest developments in goat farming. Participating in events and networks dedicated to goats provides a platform for learning, collaboration, and fostering a sense of community that transcends geographical boundaries. Whether through local gatherings, regional conferences, or online platforms, caretakers can access a wealth of resources, build meaningful relationships, and contribute to goat farming practices' collective growth and advancement.

Local and regional events are pivotal in bringing together goat enthusiasts and professionals within a specific geographic area. These events often include agricultural fairs, breed-specific shows, and community gatherings where caretakers can showcase their goats, exchange experiences, and engage with local experts. Attending these events provides an opportunity to observe different breeds, learn about diverse management practices, and gain insights into regional challenges and solutions. Moreover, it offers a chance to establish connections with nearby goat keepers, forming a supportive network that can provide guidance and assistance.

National and international conferences on goat farming create platforms for caretakers to delve deeper into industry trends, scientific advancements, and best practices. These conferences often feature expert speakers, workshops, and panel discussions covering various topics, from goat health and nutrition to breeding techniques and sustainable farming practices. Participation in such conferences allows caretakers to expand their knowledge base, stay informed about the latest research, and connect with professionals and researchers who can provide valuable insights. The networking opportunities at these events open doors to collaborative projects, mentorship possibilities, and a broader perspective on the global landscape of goat farming.

Online forums and social media groups have become integral components of the goat community, providing virtual spaces for caretakers to connect, share information, and seek advice. Platforms like forums, Facebook groups, and dedicated websites cater to diverse aspects of goat farming, allowing members to discuss specific breeds, address health concerns, and exchange practical tips. These online communities enable caretakers to connect with individuals from different regions, climates, and farming backgrounds, fostering a diverse and inclusive network that transcends traditional boundaries. Engaging in these digital spaces also facilitates real-time problem-solving, as members can seek input on issues they encounter in their day-to-day goat management.

Specialized organizations and associations dedicated to goat farming are crucial in providing caretakers with resources, support, and advocacy. Joining such organizations connects caretakers with a broader community of professionals and enthusiasts who share a common passion for goats. These associations often offer educational programs, research initiatives, and access to industry experts, empowering caretakers with the knowledge and tools needed to enhance their goat farming practices. Additionally, membership in these organizations may provide caretakers with opportunities to participate in breed improvement programs, access exclusive publications, and stay informed about policy developments affecting the goat industry.

Participating in goat-related competitions and shows showcases caretakers' efforts and accomplishments and allows them to benchmark their practices against industry standards. Competitions may include events like conformation shows, milking competitions, or agility trials, where goats are evaluated based on various criteria. These events give caretakers valuable feedback from experienced judges, helping them refine their breeding programs and management practices. Furthermore, participating in competitions fosters a sense of camaraderie among caretakers as they celebrate each other's achievements and learn from shared experiences.

Engaging with educational institutions and extension services dedicated to agriculture and animal science provides caretakers access to research-based information and expert guidance. Universities and agricultural extension offices often organize workshops, seminars, and training sessions on goat farming. These opportunities allow caretakers to stay updated on the latest research findings, learn about emerging technologies, and receive hands-on training in herd health management, nutrition, and reproduction. Collaborating with academic institutions also encourages the exchange of practical knowledge between researchers and caretakers, bridging the gap between theory and on-farm application.

Caretakers can contribute to the goat community by sharing their experiences, insights, and expertise. This may involve writing articles for agricultural publications, creating online tutorials, or presenting at local events and conferences. By sharing practical knowledge gained through hands-on experience, caretakers contribute to the collective wisdom of the goat community and inspire others on their farming journeys. The exchange of ideas within the community helps address common challenges, celebrate successes, and build a reservoir of knowledge that benefits caretakers at all levels of experience.

Mentorship programs within the goat community create valuable opportunities for caretakers to learn from experienced individuals and receive personalized guidance. Establishing mentor-mentee relationships allows less-experienced caretakers to benefit from the insights and wisdom of those who have successfully navigated the challenges of goat farming. Mentorship programs may be formalized through organizations or informally arranged within local communities. The sharing of knowledge, skills, and practical tips through mentorship strengthens the fabric of the goat community, fostering a tradition of passing down expertise from one generation of caretakers to the next.

In conclusion, joining the goat community through events and networks is a multifaceted and enriching journey that offers caretakers numerous avenues for learning, collaboration, and personal growth. Local gatherings, regional conferences, online forums, and specialized organizations contribute to the goat community's vibrant tapestry. Engaging with these platforms provides caretakers with access to valuable resources and knowledge and allows them to actively contribute to the collective wisdom of the goat farming community. By connecting with fellow enthusiasts, seeking mentorship, and participating in events, caretakers can build a strong support network that enhances their skills, enriches their experiences, and contributes to the sustainable and thriving future of goat farming.

CONCLUSION

In conclusion, "Goat Guru: Navigating the Basics of Raising Happy Hooves - A Comprehensive Manual for Novice Herdsmen" is an invaluable guide for individuals embarking on the rewarding journey of goat farming. Throughout the e-book, we have explored the fascinating world of goats, delving into their diverse breeds, behavioral intricacies, and the essentials of providing them with a suitable environment. From planning and managing a goat farm to understanding their nutritional requirements, the e-book equips novice herders with the foundational knowledge to establish and maintain a thriving goat herd.

The comprehensive coverage extends beyond the basics, encompassing essential topics such as goat health, reproduction, and practical considerations in day-to-day management. Each section imparts knowledge and emphasizes the importance of proactive care, sustainable practices, and a holistic approach to goat farming. Including real-world insights, case studies, and practical tips enhances the practical applicability of the information provided, making it a go-to resource for novice herders seeking guidance on every aspect of goat farming.

Moreover, the e-book recognizes the evolving landscape of goat farming by incorporating sections on emerging technologies, sustainable practices, and the importance of community engagement. It emphasizes the need for caretakers to stay informed, embrace innovation, and actively participate in the larger goat community to ensure the success and sustainability of their farming endeavors.

"Goat Guru" goes beyond being a mere instructional manual; it serves as a companion for those passionate about raising goats responsibly. Whether readers are interested in dairy goat farming, meat production, or simply keeping goats as companions, the e-book provides a holistic perspective that considers the welfare of the animals, the environment, and the caretakers themselves.

Ultimately, "Goat Guru" is more than just a guide; it is an invitation to join a community of like-minded individuals who share a common appreciation for these remarkable animals. As novice herders navigate the pages of this e-book, they embark on a journey filled with insights, practical wisdom, and the joy of nurturing happy hooves.

Thank you for buying and reading/ listening to our book. If you found this book useful/ helpful please take a few minutes and leave a review on the platform where you purchased our book. Your feedback matters greatly to us.

9 798886 914707 3